"Amidst the tips, anecdotes and some light hearted reflections, there is a serious message and guide in this book. The need to think through in advance how to keep yourself, your family, your business, your home, your boat and what you care about as safe as possible."

— **Gus Jaspert, Governor**

"Claire Hunter was a key player in the crisis response team in Tortola. She played an essential role in laying the groundwork to allow local leadership, public services and NGO's to respond when the crisis came, but equally important was her role in preparing the public for the inevitable communication 'black zone' so that they had taken the key steps needed to protect themselves. She has written an admirable practical insider's guide based on her personal and professional experience of both the 2017 and previous crises. I am sure it will serve as a key reference for others in the future."

— **John S Duncan OBE, Governor 2014-2017**

"As my colleagues have stressed, the key advice I have to offer is for everyone, businesses and households, to prepare as effectively as possible. Nothing says this better than the DDM moto: 'It is better to prepare and prevent rather than repair and repent.'"

— **Boyd McCleary CMG, CVO, Governor 2010-2014**

"Claire, with whom I worked closely for several years in Tortola, has produced this excellent book based upon her experiences during Hurricane Irma and which if followed will save lives. Possibly yours, or a member of your family! Please read it and adapt it for your own personal family circumstances."

— **Frank J Savage CMG, LVO, OBE, Governor 1998-2002**

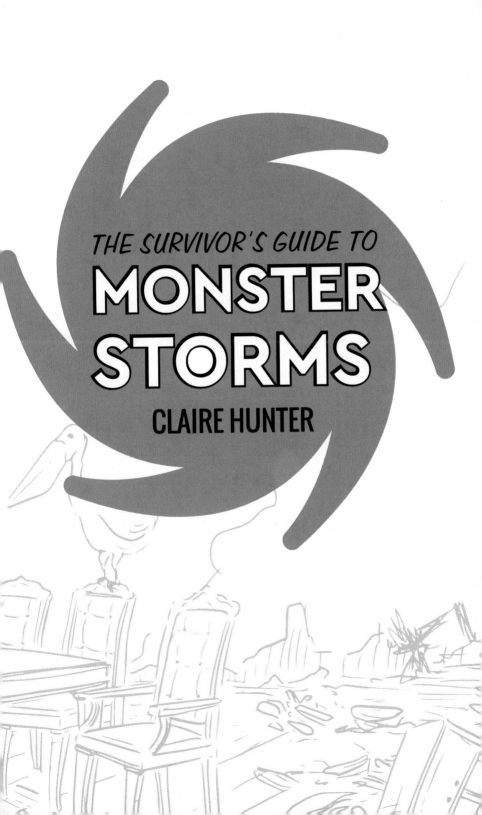

THE SURVIVOR'S GUIDE TO
MONSTER STORMS

CLAIRE HUNTER

Dedicated to the team at the Department of Disaster Management (DDM), Emergency Responders, Public Servants and the British Virgin Islands' community as a whole for their resilience and amazing spirit in the face of adversity

Published by CreatiVertical LC
154 Anchor Drive, Lake Tapwawingo MO 64015

First Edition

Authored by Claire Hunter
Creative Direction by Nick Cunha
Illustrations by Hendra Adriyasa

CONTENTS

ANNEXES

FOREWORD

Working in the field of Disaster Management means thinking constantly about the different hazards that might strike a community, how persons can be warned and the myriad ways we all need to prepare. Over the years, the most difficult part of the job has often been simply convincing persons that they need to prepare in the first place and to do so well in advance.

None of us who experienced it would ever wish to see anything like Hurricanes Irma and Maria again, but it would be remiss of us if we did not reflect on what befell us and our fragility in the face of such massive forces: enormous storm surges, record breaking winds, unprecedented rainfall. Irma was the strongest ever recorded storm to develop in the open Atlantic Ocean, even setting off the sensors designed to detect earthquakes. After she passed, our little community of some 30,000 persons faced a sea of debris never seen before and damages to the tune of $3.6 billion.

Thanks (but no thanks!) to the catastrophic Hurricane Irma, the British Virgin Islands got up close and personal with a true disaster. People who before swore that such devastation couldn't or wouldn't happen here, have had to accept that it can, because it did.

Claire has spent a significant amount of time putting so much self-reflection and practicality into this book, which is filled with her characteristic thoughtfulness and attention to detail, and contains a dash of her humor as well. This guide will be a truly valuable layman's resource for everything from gathering supplies ahead of an expected impact to pooling resources after the fact and the value of a strong network of friends, family and neighbours.

As you read, I hope you will take Claire's tips to heart, but more so I hope that you will carry with you her broader message: we are not helpless even in the face of such forces if we choose to prepare for them in advance.

SHARLEEN DABREO LETTSOME, MBE
Director, Department of Disaster Management (1998-2019)

PREFACE

The idea for a practical guide began as a 'shoulda coulda woulda' list in the first few days after Category 5 Hurricane Irma took a direct hit on us in the British Virgin Islands (BVI) on 6 September 2017. Many in our community were unprepared for the devastation and hardship that followed. I had written very thorough preparedness guides for my colleagues, and had utilised most of the resources available online, but there were still items I wish we had had, wish we had had more of, or things we wish we had done differently.

A quick search on the internet for 'hurricane preparedness guides' reveals over eight million results. Some of it is insufficient for a storm of Category 5 magnitude and much written or shared by those who have never directly experienced one, let alone survived it, in a small Caribbean territory taking a direct hit. When you're on the countdown to a major storm, who has time to sit online to determine what you need?

Hurricane Irma had sustained winds of 185 mph with higher gusts up to 215 mph (the equivalent of a superfast bullet train) with the main island of Tortola (21.5 square miles) and two of the most populated sister islands (Virgin Gorda and Jost van Dyke) experiencing the full force. We had widespread damage to the local infrastructure including roads, homes, schools, communications, water and power – my friend and colleague didn't have her own power restored for over 182 days – a few lives were tragically lost and people were displaced from their homes. Irma was so destructive she stripped the bark off trees and the once lush landscape turned brown due to massive defoliation. If the impact of Hurricane Irma wasn't bad enough in September 2017, we also had the worry of Category 4 Hurricane Jose tracking to us on a similar path (which thankfully missed us) and, two weeks later, Category 5 Hurricane Maria.

At the time of finalising this guide in September 2019, Hurricane Dorian had formed in the Atlantic, passing over us in the BVI in a disorganised fashion, before slamming into parts of the Bahamas as a catastrophic category 5 hurricane taking 40 hours to exit. As terrifying as Hurricane Irma was for us in 2017, she was thankfully a fast moving storm that hit in daylight hours. I cannot comprehend the horror of having had Hurricane Irma sit on us for such a prolonged period with much of it in darkness. Similar to our experience in 2017, Typhoon Hagibis hit the Kanto region of Japan in October 2019 as the strongest typhoon on record in over 60 years only a month after Typhoon Faxai had already caused widespread destruction. Thanks to man-made climate change and warmer oceans, what were once rare occurrences are now the new generation of monster storms.

Although we know these storms as hurricanes, they do occur all over the world with a different name depending on the region you are in. Cyclones, hurricanes and typhoons are all the same weather event and they all require the same preparedness measures. This guide is not designed to be a comprehensive manual and it will not cover every eventuality because everybody's experience will be different. It draws on existing resources and is balanced against my own perspective of the 2017 hurricanes, as well as those of friends and other professionals who remained in the

BVI post disaster. This guide now includes some additional tips from individuals in the Bahamas. Consider this 'cliff notes' to help you focus on the areas you may need to contemplate in the build-up to, and recovery from, a major tropical cyclone and I will leave it to the reader to conduct their own due diligence.

PART 1

PREPARE

I have a Dilbert cartoon on my desk which says "Our Disaster Recovery Plan goes something like this: Help! Help!"

I believe in preparing for the worst, and hoping for the best, but common sense must prevail and you have to be prepared to ultimately think on your feet. Regardless of whether you live in a potential evacuation zone or not, your extreme storm plan must include a trigger for deciding when or if to leave your home or the area entirely. A category 1 hurricane is still dangerous; slow moving storms can cause extreme damage from heavy rainfall and sustained winds; and storms can quickly shift direction or increase in intensity.

There is an old adage: "run from the water, hide from the wind." If you live in a coastal area could you be caught out by storm surge or trapped indefinitely by significant flooding? Could your home withstand sustained winds over 100 mph?

Listen to your local authorities and if you can anticipate the worst case scenario for your circumstances, you may have a much easier time in recovery.

WHAT IS A TROPICAL CYCLONE?

Tropical cyclones, hurricanes and typhoons are all the same weather phenomenon but they have a different name depending on where you are in the world.

The National Oceanic and Atmospheric Administration (NOAA) defines a tropical cyclone as a "rotating, organized system of clouds and thunderstorms that originates over tropical or subtropical waters and has closed, low-level circulation … the main parts of a tropical cyclone are the rainbands, the eye, and the eyewall. Air spirals in toward the center in a counter-clockwise pattern in the northern hemisphere (clockwise in the southern hemisphere), and out the top in the opposite direction. In the very center of the storm, air sinks, forming an "eye" that is mostly cloud-free."

Once winds reach a specific speed they are upgraded to an intense hurricane or typhoon and a category assigned to the disturbance based on the maximum sustained wind force and the scale of potential damage they can cause.

Inside the cyclone, the barometric pressure (millibars) at the ocean's surface drop to extremely low levels. The lower the pressure, the more intense the storm. The Saffir-Simpson hurricane wind scale ranges from greater than 980 millibars (causing very little damage) to a Category 5 hurricane with a barometric pressure of less than 920 millibars (catastrophic damage). When Hurricane Irma made landfall in the BVI, the barometric pressure sank to 915 millibars. This exploded doors and windows.

The strongest winds in a northern hemisphere hurricane are located in the eyewall and the right front quadrant of the hurricane. In the southern hemisphere, the strongest winds are to the left of the eye. As cyclones rotate anticlockwise in the northern hemisphere, you will typically get a northerly wind as it approaches, swapping to a southerly wind after the passage of the eye.

The direction of winds is unpredictable because the wind direction depends on where you are located under the centre. The most severe winds will generally be confined to a small area around the outside of the eye, but the outer bands of a significant cyclone can generate the same effect. It is quite often the stronger gusts within the system that create the most damage to infrastructure or areas where winds can be funnelled by the landscape such as between buildings or in valleys.

Many people don't realise that intense tropical cyclones making landfall often produce severe tornado outbreaks. In the BVI, multiple tornadoes were confirmed causing further structural damage to homes and knocking down trees and power lines. NOAA again has the clearest explanation as to why:

"Tropical cyclones spawn tornadoes when certain instability and vertical shear criteria are met, in a manner similar to other tornado-producing systems. However, in tropical cyclones, the vertical structure of the atmosphere differs somewhat from that most often seen in midlatitude systems. In particular, most of the thermal instability is found near or below 10,000 feet altitude, in contrast to midlatitude systems, where the instability maximizes typically above 20,000 feet. Because the instability in TC's is focussed at low altitudes, the storm cells tend to be smaller and shallower than those usually found in most severe midlatitude systems. But because the vertical shear in TC's is also very strong at low altitudes, the combination of instability and shear can become favorable for the production of small supercell storms, which have an enhanced likelihood of spawning tornadoes compared to ordinary thunderstorm cells."

The strong winds are not necessarily the deadliest part. Storm surges (approximately 15 feet was noted in the North Sound of Virgin Gorda) often result in coastal flooding that can cause drowning and the collapse of infrastructure. Hillsides, especially in mountainous areas, may be vulnerable to landslides as a result of heavy rains, loss of vegetation and loss of stability due to construction or agricultural practices.

A friend commented to me recently that although she had heeded the advice to not go out in the eye of Irma she never knew why and felt silly asking. The centre of the storm (the 'eye') is an area of calm weather which can give a false sense of security as more tropical force winds will return rapidly leaving insufficient time to get to safety. More people are killed or hurt during this period trying to make repairs or having a better look. You should stay indoors in your safe area until you are alerted to the 'all clear'. You may have no choice but to use this brief window to make quick emergency repairs or to get to another (safer) location but it's not recommended.

TRACKING

During the official hurricane season (June to November) I monitor multiple weather sites, including our local Department of Disaster Management, to stay aware of any inclement weather. I also keep a close ear to the sailing community: when they panic, I panic! Learn your local resources as well as the warning systems.

Once a system has been identified, meteorologists will begin forecasting the projected path of the system using computer models, satellite data and other weather variables. This is commonly known as the 'cone of uncertainty' as the exact path is not known and neither is its size. As it can change drastically at any time if you are in the cone, or even just outside it, don't wait until an official warning to begin preparations.

THE ESSENTIALS

Food, water and shelter are your most basic needs. And the same goes for your animals. In addition to your own supplies, prepare on at least two to three weeks of food and sufficient, safe, drinking water for you and your pets. I under-estimated what I needed but thankfully I was able to pool resources when I moved in with friends.

SHELTER & THE BACK UP PLAN

Take a hard look at where you plan to ride out the storm and either identify your 'safe room' in advance or find an alternative location especially if you are flood or landslide prone, or there is a high likelihood that you could be ordered to evacuate. Remember that winds are stronger at higher elevations, so you should bear this in mind if you live in a high-rise apartment building or a hillside property. No power likely means no elevator! If you live in a heavily wooded or remote area, fallen trees could block you in for days or until someone comes to find you. Options include friends, relatives, emergency shelters or out of town.

Get off your boat and, if you can, get it out of the water to hard storage. You just have to look at one photo of the devastation in the marine industry in the BVI to realise it is not a safe decision to ride out any cyclone on your vessel.

If you decide to stay at home, identify an area in your home where you will ride out the storm away from windows and doors. This will become your 'safe room'. It is best to use an interior hallway/corridor or a windowless bathroom on the lowest level in a concrete building and away from flood prone areas. I have friends who took refuge under the stairs and bathrooms became the room of choice for many. As I had an open plan living area on the ground floor I chose the kitchen as the windows were shuttered, we were sheltered by tall built in cabinets and we were sufficiently away from the front door if it blew in (it did). Many friends additionally relocated a mattress into their 'safe room' to shield them from any debris.

An interior room on a lower level is also the safest place to be in the event of a tornado. If you can, get as many walls between you and outside as possible.

Is your 'safe room' safe? I spoke to someone who discovered too late that he had forgotten to reinstall the door to an internal room.

When you review your home with safety in mind, determine how you might get out of it at short notice and identify any possible escape routes in advance. If you have only one potential exit from the building you may wish to reconsider staying there: if a tree were to block your only door preventing you from getting out, this could be a major problem.

Pre-position at least two days' worth of non-perishable food, first aid kit, baby supplies and water for each member of the household in your 'safe room'. It is also a good idea to have some essential tools with you in case you have to do emergency repairs or make an emergency exit. I spoke to one person who had boarded up his house but couldn't get out because part of the structure had collapsed. A neighbour came to help with power tools and removed some of the boards to help them leave.

Learn your community evacuation routes and emergency shelter locations. Where would you go? And how would you get there especially in storm surge, high volumes of traffic or barricaded roads?

Reach out to any friends, colleagues or neighbours who live alone or who live in vulnerable properties and invite them to join you. Insist that they bring their own supply of food and water. It is not advisable to ride out a cyclone alone. More the merrier, if only for safety.

Be aware that most emergency shelters will not allow pets and rescue officials may not allow you to take your pets if you need to be rescued. Call hotels/friends/family members/vets/boarding facilities in a safe location in advance to arrange foster care for your animals or to ask if they would accept pets. Do not abandon them.

WATER

Drinking Supplies — You can survive without food for about three weeks, but you can't survive without water for longer than three to four days. Plan on at least two to three weeks' worth of drinking water for each member of your household and include your pets. The World Health Organisation recommend at least two litres of drinking water per person per day which is approximately half a gallon per person per day. I would recommend doubling that especially if you live in a hot climate. You should anticipate heavy physical work post disaster, walking longer distances than you perhaps are used to, and managing potential injuries. Young children, nursing mothers and the elderly may need more. If you plan on using powdered milk or baby formula, you will need to incorporate more water.

I cannot drink the water from my taps and, like most in the community, am in the habit of refilling at the water stations across the Territory. I have three five-gallon jugs which will be re-filled at the first weather alert which will be the bare minimum to sustain myself and my two dogs. I also keep empty one-gallon water jugs and these would be filled with tap water for general use also at the weather alert. I can then use it for cooking, cleaning, bathing or boiling.

Store drinking water in unbreakable food safe containers such as empty bottles or gallon jugs and clearly mark what is for drinking and what is not, especially if you have children in the home. Store bought water (single use) is the easiest option for storage but the most expensive.

Rotate any stored water every six months to remain fresh.

General Use & Storage — Use every available clean container to then store extra for general use, bathing and cooking: fill the bathtub (if you have one), washing machine, empty bottles, rubbish/trash bins, and empty coolers.

If you have space, consider storing as many buckets of clean water as possible to help with cleaning up post storm.

Laundry will become a luxury if you lose power after the storm so consider getting to those loads of dirty clothes you have been putting off before the storm hits. Also make sure you have liquid detergent so you can hand wash after the storm.

Hand sanitiser and baby/wet wipes will save on precious water. And buy extra.

Water Source — Think ahead to your water source options post storm. If, like many of us in the Caribbean, you rely on a rainwater cistern remember that water pumps need a power source and you will need a plan to access your main cistern if you have no backup source of power. Cistern lids are likely very heavy or difficult

to access. You can use a bucket and a rope or Ashton, a handyman I know, said he left a hosepipe permanently in his cistern outlet after the storm having previously syphoned out the air and then kinked the hose.

If you do have a cistern, have some form of cleaning agent available such as chlorine tabs or liquid bleach that you can use after the storm to clean the water supply. Homeowners can consider installing valves on the downpipes to prevent muddy water during the storm from getting into the cistern and contaminating the entire water supply.

Other water sources for general use include swimming pools, the sea, ponds and lakes.

The marina where I lived had their own water desalination plant for the resort and the new build condos which was initially damaged after the Hurricane. Once it was repaired, they realised that many in the community had no access to a clean water source and extended a pipe to the main road at no cost to members of the public. They left this line available for approximately eight months.

FOOD

The supermarkets will reopen after the storm eventually but, based on the BVI's experience they opened with limited hours and the queues to get into one of them initially were very long. They were also cash only for some time as the loss of power prevented the use of credit card machines. One of my new housemates had the flexibility to get in the long lines at the store and evenings were exciting seeing what she had managed to find. The day she came back with some fresh fruit was a good day and finding alcohol and cigarettes a little like Christmas coming early.

Panic shopping at the first weather alert will empty supermarket shelves fast. If you have the budget, either set aside time at the beginning of the season to shop for disaster supplies or gradually build your stock up, perhaps buying one extra can or extra water when you do your regular shop. Remember to rotate and replenish your stocks and keep it separate to your normal pantry items. Replace stored food every six months to keep it fresh.

Prepare for at least two to three weeks' worth of non-perishable food for each member of the household. Remember that items in your freezer will only last 48 hours once the power goes, assuming you stop opening the freezer door, and perhaps 24 hours in your refrigerator so don't stock up on perishable items. And when buying food think ahead to the meal combinations that you may need to prepare, dietary needs, and how you will cook it. Canned beans and pulses are great sources of protein and fibre, pasta and potatoes require precious water to cook and tins do have a shelf life.

I have many friends who now have MREs (Meals Ready-to-Eat) in their contingency supplies for good reason as they have an extremely long shelf life. I think we're still using up the discarded sugar and coffee packets in the office as we had the Royal Marines of 40 Commando working out of our office for some time. Spare biscuits were generally fed to the Governor's dog who became 'Head of Morale'. Meal replacement drinks are also an option if you can get past the sweetness and you have some way of getting them very cold.

Fresh fruit is generally too perishable to store but if you would prefer to have

SANDWICHES ARE A HEALTHY MEAL OPTION

some fresh items in your contingency stock, consider buying apples, oranges or pears. They're good sources of vitamin C, keep longer than most fruit, and will help keep you hydrated. Best to buy them, if you can, a few days out from storm impact and try to purchase slightly under ripe.

Unless you plan to use uncooked rice as a drying agent for a wet phone, it is not recommended for contingency supplies as it does not store well. But do buy it in bulk if you have animals, especially livestock/horses, as you can bulk out their food supply.

Don't forget to buy comfort foods to relieve stress and chronic boredom.

If you're addicted to your electric coffee machine – and instant granules simply won't do! – invest in an alternative coffee maker such as a cafetiere so long as you have a method of boiling water. You can also take out the filter cup of an electric coffee machine and run water directly through it into a cup or jug.

THE NECESSITIES

GRAB BAGS

You may only get minutes to leave, so consider packing a small grab bag for each member of the household including passports, spare phone and call credit, portable chargers, radios, flashlight, pet paperwork, insurance, change of clothing, medicines, travel toiletries and medical information. Make sure the grab bags are in the 'safe room' and ideally attached to you.

Alternatively, a waterproof neck pouch/wallet can be used to safely store your phone, passport and car keys or attach your passport to your body in a waterproof bag. One piece of sensible advice from an individual in The Bahamas following Hurricane Dorian was to put your passport, along with any and all available cash and insurance policy numbers, in two ziplok bags and tape it to your body.

I had a grab bag permanently with me which included a work satellite phone, passport, cash (including Sterling in case I had to leave), a change of clothes, and important papers – including the paw prints of my beloved dog Lucy who had only just passed away. The items you pack are what are most important or useful for you.

Explain the contents of the grab bags to your children, why you have the items and how to operate them.

EVACUATION

If you are ordered to evacuate – GO! However, if you live in a flood prone area and leaving is not an option, for whatever reason, consider purchasing a personal floatation device with a strobe light for each member of the household, including your pets. My colleague insisted on having lifejackets available for her children when their original choice of safe location was near sea level.

Leave enough time to avoid being trapped by severe weather and let others know where you are going. You can use technology such as geofencing by notifying your location to family outside of the area and keeping them updated as you move. I injured myself trying to find a neighbour immediately after the hurricane had passed only to discover that my friends had left the property pre storm. Despite an effective flashlight (and an argument with my then boyfriend about going out at all), I did not anticipate the volume of debris outside in the dark and tripped over a metal table and chairs that a neighbour had not secured!

Be sure you and your family are well fed before going to the home of a friend or a public shelter. My hurricane buddies thoughtfully brought a pre-cooked lasagne which gave us our only proper meal for 48 hours.

Before you go, anticipate power failure or contaminated water sources when you return and fill as many containers as possible with water for drinking and general use. The bathtub and washing machine (if you have one) are good options.

Shut off utilities at their main switches. If you have propane gas, shut off the valve, disconnect the bottle and anchor securely. Inside your property might be the most sensible option.

Protect windows, lock windows and doors and ensure your outside areas are clear of any furniture, rubbish bins, or other items that could cause damage in high winds.

EMERGENCY SHELTERS

Emergency shelters will normally provide no sleeping or food facilities (you generally need to take your own supplies) and will not normally accept pets. Plan on meals that do not require heating as cooking options are not guaranteed and remember that the food is intended for post storm so you may wish to take enough for a few days.

CHILDREN

I spoke with various parents who went through Hurricane Irma who gave the following advice for preparing, and reassuring, children throughout what might be a terrifying experience.

In an age-appropriate way, be honest about what might happen, noises they may expect to hear and any evacuation plan so that they know what to expect. This should include the contents of the grab bags. My colleague explained these items to her children which included training on UHF radios. She wanted to empower her children to be able to act in the unlikely event something happened to her or their father.

Reassure your children about their safety and manage your own anxiety which will quickly transmit to others.

If a friend or family member can take care of your children, especially very young ones, so that you can get properly prepared in advance, this is recommended. Preparations will take longer than you anticipate. Your children can also help with some of the preparations by helping to bring inside any toys or bicycles kept outside.

One parent I know insisted her kids wore their bicycle helmets during the storm and had lifejackets available when their original choice of safe location was near sea level.

Make sure your children, as well as others in the household, are wearing proper shoes during the storm and have some warmer clothes available – shock can make you feel very cold.

One parent thought ahead and made sufficient sandwiches to last a few days. The sandwiches were nutritious for dinner if they had no ability to cook; eating relieved boredom and stress; and they will keep fresh for a short time.

Think ahead to entertainment: my good friend invested in good quality headphones and her three-year old daughter spent the entirety of the hurricane watching movies on an iPad seemingly clueless about what was going on outside. This didn't work for every family however and games and singing was needed to try to distract frightened children.

ANIMALS

Be a responsible owner for any animal under your care. Have a basic first aid kit containing antibiotics, anti-inflammatories, painkillers, anti-parasite medication, bandages, 'pee pads' and lots of paper towels. Typical ailments will be cuts and lacerations (especially lower limb injuries in loose livestock), respiratory infections and fungus/skin infections.

Domestic & Small Animals — spay or neuter your pets. This brings major health and social benefits as well as helping limit overpopulation of unwanted dogs and cats in the aftermath of a major storm with fences damaged or destroyed. If cost is an issue, identify if there are any spay/neuter programmes in your area to reduce or eliminate surgical costs.

Be prepared all year and not just during a cyclone season, especially if you are likely to want to make an international move with your pets after the storm. Make sure your pets are up to date with their inoculations, that you have an accurate and recent photograph, and that your pets are microchipped. Ask your vet for a copy of the rabies certificate when you do their annual check-up. The rabies and microchip certificates are the basic documents for international travel. In advance of a named storm, make contact with your vet to check the status of your pet and query any paperwork.

If your domestic animals live outside, bring them indoors well in advance of the storm. If they cannot come into your home, do not leave them tied up so that they can defend themselves against flooded areas or flying debris. I hate to think how many animals died this way in the BVI during hurricanes Irma and Maria.

If you know you will evacuate to another location, take your pets with you or make alternative arrangements for them. If it isn't safe for you, it isn't safe for them.

Local vets and their inventories may be affected by the storm. Ensure you have at least three to four months' supply of medications in stock, including tick prevention, more if your pets have a chronic condition. The lack of foliage, loose livestock and fast-growing grass in the BVI post Irma created a tick epidemic with domestic animals becoming the host of choice.

Keep pet food dry and secure in watertight containers. Tinned food has a longer shelf life and can help hydrate your pets. If you run out, you can use human-grade canned meat such as Spam as a backup.

Ensure you have a carrier for your pets especially if you have small animals such as guinea pigs and rabbits. I had no crates, so my two dogs, and my friend's dog, were kept on lead and they were attached to a human throughout the storm. One of the marina cats was brought inside for safekeeping and she spent the storm in a borrowed cat carrier (which I still need to replace). It did take a few cans of food to get the greedy cat into her bag, and she bit my neighbour who was trying to help, but the bag quickly became her friend when the hurricane came through.

If travelling internationally after the storm is a realistic possibility, remember that pet crates will need to meet international airline standards. It is beneficial to attach an up to date photo plus your pet's name to the top of the crate for identification purposes. If there are multiple kennels, for example at an airport and the animal escapes, this has the added benefit of officials knowing who to look for!

Discuss anxiety medications with your vet. I have a friend who swears by

'Thundershirts' and I have used homeopathic treatments in the past on my old dog who was very sensitive to bad weather. Treats or a favourite toy will reassure.

If you have absolutely no choice but to leave your pets behind in your home alone, leave toilet seat lids up so that they have a water source (but don't use bleach or cleaning tablets in your toilet tank), leave a mattress on the floor where they can bury and protect themselves from debris, and if flood prone something that they can climb onto to escape rising water. Leave a notice too that animals are inside. This can be a pre-made poster providing some basic information about your pets including medical needs and temperament or you can use spray paint on the outside of the property. Anything to leave the message.

Domestic birds — Before Irma hit, my friend had crated up every cat, tortoise, guinea pig and rabbit in her farm and left them in the downstairs area of her home where they would be safe. As she says, a lot of crates! But she also had two parrots in their heavy cages in the upstairs apartment which you can barely move let alone run with. She said not everyone's Parrot or Budgie likes to be handled. Her Macaw does not and his beak can break fingers. She had no time to worry about broken fingers during the eye when she found two empty cat carriers and had seconds to transfer the birds. The message here is that birds need to be in plastic or wire crates like the smaller domestic animals. And you need to transfer them well before storm impact so that they are not freaking out when you are belatedly remembering to manhandle them into a tiny box! After the storm she acquired two Budgies from one of the international schools but their cage was ridiculously flimsy and she has no idea how she managed to keep them from her cats for days afterwards. The Parrots were fine in the cat crates and, as soon as she could, she built them a perch inside but she said she should have done all of this the day before.

Livestock/Farm Animals — If you cannot evacuate them, don't secure livestock or horses. If you confine them they could be seriously hurt, or even killed, by collapsing structures and flying debris that may have been avoided outside.

Horses are apparently like boats: they face the storm 'head to wind or rain' until it makes their eyes stream and then they turn 'arse to wind or rain' – tails ending up between their legs - and when that's too much they will take off to seek shelter elsewhere. The rescued horses at a friend's farm all had lower limb injuries from debris. The private racehorses that had been left in their stables at the track all sustained upper limb and head injuries from exploding steel, wood and nails. Those that had been left in the open, whilst injured, survived.

The experts have recommended that most livestock will instinctively seek shelter in trees and tall brush. The best pasture has no non-native trees that would uproot easily, no barbed wire fencing, no overhead power lines or poles and is at least an acre of open space.

Keep food dry and identify a secure area to store hay and sweet feed. If there is a chance that you could run out, consider buying rice in bulk in advance of the weather (whole wheat is best) and you can bulk out their feed to extend their food. Our vet advises that you can take two days' of feed and extend it to six potentially. Animals, especially livestock, must have fuel for energy.

Consider your water source post storm. External water tanks may be destroyed so try to store a spare securely. You will also need to keep your water source clean and chlorine tabs can help sterilise.

Think ahead to methods of identification so that you can be reunited with your animals post disaster.

PREPARING YOUR HOME

I have a friend who has goats. She built them a shed in her garden and spent the entirety of the hurricane petrified that she had committed them to a certain death. The goats were fine as was their 'Category 7 shed', minus two (yes only two) screws. My friend believes the success was down to the sheer volume of screws she used to build the shed and some lady luck.

I saw many sad sights in the days following Irma's impact. I can still remember seeing the tragic sight of homes that had been boarded up suitably, but the roof had been ripped off by the wind, or solid concrete structures reduced to rubble. This included our Department of Disaster Management and the National Emergency Operations Centre. They were forced to make a terrifying evacuation from the building during the hurricane to a neighbouring property and then re-established operations in a conference room at the local hospital as soon as the hurricane had passed.

There are many engineering and construction studies being done to determine which home designs, construction methods, and pitch of roof (30-40 degrees) can better resist storm force winds and what remedial actions homeowners might be able to take to avoid total structural failure when extreme winds or debris penetrate the building. A few friends swear that their roof overhang contributed to their roof failure when it peeled up under the sheer force of the wind and there were many cases of the vibrations from the hurricane causing bolts and nuts to undo – engineers are now looking at lock nuts. Shorter overhangs are better, but all timber needs to be strapped down. There is no point in strapping down the eaves of the rafters if the ridge is not strapped and we saw many roofs which had peeled open from the apex of the roof. But check first with a structural engineer and an architect before you make any significant design changes to your home or schedule a building inspection if you have any concerns about the building's structural integrity.

At the minimum, homeowners should check for excessive rust, invest in storm shutters, reinforce external doors and perform preventative maintenance on their property including the roof.

Protecting your roof — The highest forces on a roof are at the apex, the eaves and along the line of the hips (if you have them). The roof sheeting and boarding should be more frequently fixed at these locations. Ridges and hips must have metal straps over the rafters and should be clipped or cast at beam – most roofs in the BVI were cast at beam but not strapped at ridge.

Ring shank (or improved) nails provide better pull out resistance, but the total number and the penetration of the nails into the main timber members is important.

Asphalt shingles are rated from A to H, with A being designed for up to 40 mph, and H up to about 120 mph. The vast majority of the shingles in the BVI were the lowest specification.

Standing seam roof or metal roofs are good if fixed down sufficiently. The felt 'crinkly tin' shaped sheeting and asbestos roofs are not good.

One structural engineer recommended the use of 3" timber, as there is more meat to fix down to and they are less likely to fail sidewards by buckling.

Protecting Windows & Doors — It's surreal listening to the sound of neighbours banging in plywood to their doors and windows when the community is on count-down to impact.

Do not tape across windows as this will not stop them shattering and could create an even bigger danger as the shards of glass will be bigger. Instead homeowners should invest in some impact resistant windows or shutters or install 5/8" exterior grade or marine plywood across the windows to prevent wind pressure from building up in your home and potentially lifting the roof off your property. Formply is a cheaper alternative to marine grade plywood.

It is best to measure and purchase the plywood well before any weather warnings as the stores will quickly run out. Consider marking each panel for the relevant window and anchor using masonry screws – Tapcon 3 ½" concrete screws have been recommended. You may not have as much time as you think you need to attach them. I know one person who recommended fitting the plywood to the actual window frame, rather than across it, so that wind could not get behind the frame, but this will only work provided the frame is sufficiently well fixed into the substrate.

If you purchase impact-resistant or hurricane rated windows or doors understand that nothing is really hurricane proof: window frames may blow out, impact resistant glass can still crack and calculations may need to take account of your topography as wind accelerates over the top of a hill - no hurricane glass will be able to resist an object at 150 mph. One structural engineer said they are trying to encourage their clients to use secondary protection against impact damage. It is pretty terrifying when your windows are slamming open and shut and glass flying off with every impact.

There is a healthy debate going on about hurricane shutters and whether roll down or accordion models are better. All I can say is that the window shutters on one side of my condo were accordion and they held well. Others didn't lock so well due to age and corrosion. They do need regular maintenance with WD40 and don't leave it until the official warning to figure out how to close them. It turns out I cannot do them alone and I needed a strong friend to climb a ladder and hang off some beams to get them closed as Hurricane Irma began to make her presence known.

Entry doors are easily damaged by high winds and the door hardware may fail if it has not been properly maintained. You may need to consider replacing door knobs or latches to ensure doors stay closed. Double entry or 'French doors' are particularly vulnerable. Entry doors can also be a point of entry for water even if you are not in a flood prone area due to the sheer volume of rain. A friend had to sacrifice her sofa cushions at the bottom of her doors to try to keep the water out! If you don't have protection at the windows (not recommended) and you have external mosquito screens, friends recommend you remove them as they will not be there post storm! Another weak point of homes are garage doors.

Whether you purchase speciality windows and doors or go the plywood route, they're only as effective as their original installation and being properly anchored to the walls of your home. I know a few people who invested in hurricane rated doors to regrettably discover the contractor or handyman had skimped on the number of recommended screws when installing them …

Electronics & White Goods — Consider spare cool boxes for storing refrigerated items, such as drinks or condiments that might be in frequent need to avoid opening the refrigerator. Use the other for storing ice but pack as many bags of ice as you can into the freezer and transfer when you need to. An alternative is to freeze large containers of water. They always keep and if you leave at least one in your freezer it can help to stabilise the freezer's temperature.

Turn the refrigerator and freezer to coldest setting at the weather watch and keep the doors closed. If the power goes, consider wrapping the refrigerator with a spare duvet for extra insulation but don't do this when powered up as the refrigerator needs to let heat out, normally in the back of the appliance. You can do the same with any cool boxes.

Unplug electronic appliances or devices such as TVs, microwaves and computers to mitigate any power surges or lightning strikes and wrap in plastic. Turn off the breaker to your air conditioning units.

Weatherproof any exterior breaker panels with duct tape to limit water intake and then turn them off before storm impact.

Utilities — Shut off utilities at their main switches. If you have propane gas, shut off the valve, disconnect the bottle and anchor securely, strapping the bottles down if necessary. Inside your property might be the most sensible option.

Outdoor Areas — Do not underestimate the sheer force of the wind. Bring in everything from outside and secure safely. This includes balconies. That plant pot or table and chairs might not affect you but could damage a neighbour's property if picked up by extreme winds. The force of the wind is proportional to the square of the wind speed. The force gets much larger very quickly.

If you can, disconnect or block your rain gutters and downspouts as you want to avoid storm contaminated water and debris entering your household water supply.

Trim trees and shrubbery, especially those that could get tangled up in electrical or telephone wires. This will assist in the recovery effort to get all systems back up and running quickly after the storm.

Tackle any slow exterior drains that could contribute to flooding. Drain clearing products can help. Remove any rainwater cistern overflow screens.

Securing Personal Effects — Generally declutter. For the many hours we sat on my kitchen floor, my hurricane buddies pointed out the items I still had on the top open shelf in my kitchen. I had also forgotten to remove my archery equipment which was still hanging on the wall and I had a mad scramble to grab my Bow and quiver full of arrows when the doors blew before one of us was injured. We never did find the iPad and I kept finding Monopoly money for weeks afterwards …

If you are flood prone move possessions from lower to higher levels of the house if you can or simply try to elevate from the floor.

Remove and protect any valuable items including vital documents, pictures, paintings, photos. I removed all my paintings from the walls, covered in large plastic bags and secured in plastic boxes in case of flooding. I also packed up all my ornaments and loose photos and secured in a plastic box. I looked like I was moving out. As it turns out I did, and did not return home for a further two months because of damage. Others never returned home at all.

I lost some very expensive area rugs because I did not anticipate the water that would enter my home and then I didn't act in time to save them post storm. Roll up any carpets and try to store away from any flood prone areas.

Don't consider your empty dishwasher waterproof. If the area floods so will any items in it. That being said, when the condo's doors blew out and items were blowing all over the place, I did grab what I could and threw it in the dishwasher for safekeeping.

Swimming Pool — Pool experts recommend not emptying your pools before a major storm to avoid oversaturating the ground, but you need to consider your own personal circumstances. I have a friend whose pool was at a higher level than their guest cottage and the overflowing water meant they had no choice but to dump the water. Another said if they had emptied the pool, they would have had less damage as a tornado landed in the swimming pool and took all the water through the house.

The Royal BVI Yacht Club had no choice but to store their sailing dinghies in the marina's now empty swimming pool when Hurricane Maria came through two weeks later. Everything had to go in the pool as there was no other secure storage left after Irma and they were very grateful for the help of some strong Royal Marines!

Consider also if there will be anyone post storm to take care of the pool as it could become a breeding nightmare for mosquitoes, frogs and other wildlife.

Secure any skimmer lids and store swimming pool furniture and tools inside your pump room or property. As a last resort they can be 'drowned' in the pool.

One local pool expert recommends shocking the pool with double your normal amount of shock chemicals at least twelve hours before storm impact and running the pump for as long as power permits. Another says save your money, walk away and deal with it afterwards!

In the BVI, the main power grid is turned off once winds reach 40 mph. If you determine it safe to do so, now is a good time to turn off the breaker to your pool power. Alternatively, you could also ask a pool professional to adapt your timer to turn off before the expected storm impact.

This is a good source of household water post storm.

PREPARING YOUR BUSINESS OR OFFICE

The same methodology for home applies to your office or business. Above all else, ensure that your staff are given sufficient time to complete activities at home. Making staff work late risks their own safety in getting home to get ready.

The more efficient your office shut down means the smoother you can get back to normal once the storm passes.

Have a business continuity plan and practice it: if your premises are damaged, identify a backup location where you will work. Which elements of your business must be operational within hours and what routine tasks can be deferred? Do your employees have the right equipment if you expect them to work remotely?

If your organisation has a plan to evacuate staff to another country, ensure affected colleagues have valid travel documents throughout the cyclone season, including visas, and make sure the plan takes account of any staff members that have pets. Animals are not normally included on private flights and staff may be reluctant to leave them.

Depending on your duty of care policy, identify where staff will be throughout the storm and who they will be with. Obtain emergency contacts for each member of staff and check the details are correct at the storm watch as family members may be travelling or otherwise unavailable. What is your plan for obtaining updates on staff welfare and notifying these emergency contacts especially if the mobile networks are down? Pre-positioning satellite phones with key staff may be the option. One friend stood outside his office every day at a specific time until all of his staff had been accounted for. Another firm plotted where their staff lived and colleagues were assigned to physically check on each other once it was safe to do so.

Business continuity also includes backing up your data and knowing how to recover your files.

Store important documents in waterproof containers or cabinets. Disconnect all electrical items, including any office refrigerators, and wrap all IT items, including UPS systems, in plastic. You should consider shutting down your servers.

If you are flood risk, you should remove all electrical items and documents from the floor and away from windows. Some firms have designated storm rooms where all vital documents and equipment are stored.

If you have satellite devices as back up, make sure they are regularly charged and tested, contain key numbers, and all staff know how to use them.

Consider obtaining extra cash from the bank – if the banks are not operating for a while you may need to make initial payments (including payroll) in cash.

Ensure you have sufficient drinking water and cleaning products for any clean up post storm.

If you have a generator, ensure it has been serviced and has been refuelled.

Identify how long all these preparedness measures take, who the responsible staff are, and plan for it.

BOATS

My colleague and her husband had been promised space in one of the official hurricane holes for their power boat but at the last minute lost the spot. In desperation they tucked the boat into some mangroves and crossed their fingers. The boats in the hurricane hole were mostly destroyed or never seen again – theirs survived minus the Bimini frame which they forgot to remove.

Preparedness is key — You are responsible for the damage your boat may inflict on others and there are normally severe fines for abandoning any vessel post disaster especially if it ends up blocking a channel. It is not fair or right to make your boat someone else's problem.

There is no way of knowing if any boat or marina could reasonably survive a storm of Irma's magnitude but my friend in the industry has the following advice:

Preparing your Boat — Know exactly what you are going to do in order to prepare for a storm, the time it will take and what supplies you will need. And do all of this before the cyclone season begins. Do as much of the preparation as you can. Leave the extra lines and fenders on the boat so that you are not running around getting things you have forgotten and wasting valuable time when a storm is going to hit.

Educate yourself about storms and what sources of information for your area are the best. Start studying them early on and think what it would mean to you if that storm came towards you and where your boat is. What will it mean if the storm passes to the South of you? What would be the difference if it passes to the North? Direct hit, wind force, rain and storm surge are all variables that you need to be aware of to protect you and your boat. The National Hurricane Centre/NOAA is a valuable source of information for those of us in the Caribbean.

If you are new to the area or new to boating, find local knowledge from as many sources as you can. Discern between good and bad by consistency and who the most credible source is. But do your research first. The more you know before getting local advice, the better you will be equipped to understand the advice and the more readily a good source is going to invest valuable time in helping you.

Remove all sources of windage. Do not leave lines slapping on the mast. Look for sources of chafe and protect all the lines.

A boat is only as safe as the ones around it — A poorly prepared boat coming loose in a storm will undermine all of your preparations. The same goes for boats stored on the hard. Check them for problems and take an honest look at your own. Do not let yours be the one that wrecks another boat.

If your boat is on a trailer — Pick a protected spot, tie the trailer down, chock the wheels well and strap the boat to the trailer. Depending on the boat and the severity of the storm you could consider filling the boat with water to give it extra weight.

If your boat is in a marina — Factor in extra time to help others. Everyone is in this together – you don't make it through these things on your own.

Use proper dock line. You will double, triple lines up given the amount of wind you expect. You can use rigging line if you must but note that it has very little stretch. For this reason it should not be tied as taut as the dock line. As the dock line stretches to its limit, that is when the rigging line should take on the load. The stretch of dock line is a shock absorber and that is why it is good. Rigging line would also be better used for the longest runs.

If you are worried about the cleats on the dock coming out, use chain to run through them and around the dock. If the cleat breaks off at least the boat won't be totally free.

Wear gloves to protect your hands. Just handling the lines for a day will leave them sore even if you are an avid sailor.

Do not depend on fenders - they will burst and your boat will be damaged. Tie your boat away from the dock. If you share a slip with someone then spring your boats towards each other and off the dock. If not, find other things or use anchors to spring away from the dock. Be able to jump to and from the boat. Wind will also blow your fenders out of place. Tie two or three of them to a car tire and then put the tire in the water. This will help hold the fenders down.

Now test your lines: the boat should barely move when you use the engine in forward and reverse. If you have help, then one can drive towards a slack line whilst the other takes up the slack. If you are setting an anchor then set it first.

Ensure your batteries are charged. Bilge pumps should be wired directly to the batteries. Turn off power to everything else including your engine(s). Most modern day engines have sophisticated, and expensive, computerized operating systems. They have a much better chance of surviving if they are cut off from electricity and many expensive outboards were destroyed because of this. The engine of a sunken boat can be brought back to life but if the computer was still energized that cost of repair goes up exponentially. Lastly, unhook from shore power.

If you plan on checking your boat during the storm — This is not advisable, but if you are expecting to check your boat during the storm then expect to have to adjust lines. Try not to use line for more than one run from dock to boat. Having to untie multiple lines in a Category 1 or 2 hurricane is hard and dangerous for you and your boat.

If you do go out, don't go alone - buddy up – and only check the boat if it is safe. Be mindful of things that will fall or break and wear a bicycle helmet. Trees are obvious sources of danger.

The lines you tied will stretch. Be prepared to tighten them back up. Having a buddy will pay dividends in time and frustration.

Please remember that your boat is a boat. You will live well without it.

Finally, if you have a dinghy fill it with water, go home and drink rum.

INSURANCE

Many home and business owners in the BVI either had no insurance or they were under insured. Whether under insuring was done intentionally to keep their premiums down or they never thought to upgrade their policies as time went on can become an expensive mistake.

It might sound obvious, but make sure existing policies are up to date and remember to get your home or business revalued regularly.

Not paying your insurance premiums does not necessarily mean you are not covered, but if you receive a cancellation notice that is different. Check with your insurance agent as every jurisdiction is different. There were many policy holders in the BVI who had lapsed on their premium and in an extremely generous demonstration of good will after Hurricane Irma, one local insurance broker honoured these policies and simply deducted the arrears from the final claim.

If you do decide to take out a new policy, bear in mind that as a storm approaches insurance companies will likely have a cut-off date for accepting new policies and there may be additional waiting periods after a storm has passed before you can file a claim. They will normally have no problem with renewing existing policies.

Home — If you hold a mortgage on your property the mortgage provider may insist that you hold building insurance as a condition of the loan. Be aware that this likely means that in the event you have to file an insurance claim for storm repairs you may have to get your mortgage holder's approval before you can cash the insurance funds to rebuild your home. The bank may also take over control of the insurance money or put the balance in an escrow account until the repairs are finished. This is to protect the bank's financial interest in the property and is standard practice. You should check the small print of your policies and speak to your mortgage provider for further information.

To avoid under insuring you should regularly revalue your home by contacting your local surveyor. You may need to increase your policy limits if the cost to repair or replace your property has gone up as you are not insuring your property for the cost you paid for it, but rather for the amount it should cost to rebuild it. Ensure you have a copy of the policy in the event you need to claim.

Tenants in rental properties should understand that their landlord's homeowners insurance (if they have it) will not cover the tenant's personal items. Contents insurance is reasonably priced or try to have a reserve to cover the costs required to replace any personal items such as your TV. Know how much your deductible will be as a small claim may not ultimately be cost effective.

Seek legal advice if your landlord insists you take out household insurance to cover their personal items if you rent a furnished unit.

Have an inventory of home contents that are being insured and photograph everything. Your inventory should include a full description of each item and receipts if available. You may need to itemise in advance any high value items and it is best to do all of this in advance of the season. You will not have time at an official warning.

Don't forget to back up this information, including any receipts for building, home improvement or purchase price, and store it safely preferably in the cloud.

Business — Preparation for hurricane season includes having property and contents insurance. In the words of a close friend who was under insured (and who reproached me for reminding him of his biggest blunder to date) said, "...don't think well, this will never happen to me". He said the insurance company gives you everything you need to be reminded of adding on new assets every year when your premium is due.

Policies should include any damage and loss to your business from severe weather, including clean-up and debris removal. The policy cover should include furniture, documents and equipment. If you cannot find an insurance policy that covers you for everything you need, you should look at a number of policies that will.

If you operate a business from home don't assume that your personal household insurance will cover you.

Marine — It is best to go with an insurance company that specialises in marine insurance to ensure you obtain the right coverage.

Contact your agent to confirm your boat is covered and make sure you understand what is covered in your policy and any associated deductibles, limits and exclusions. Most policies will be invalidated if you don't have your vessel in an official hurricane hole, some policies do not cover you if you are in the water and you may additionally be asked to unstep your mast to further minimise windage.

If you only have liability insurance check if it includes salvage and wreck removal and you may need to consider separate coverage for fuel-spills.

If you live on your boat, be upfront with the insurance company – not all policies will prohibit this – but for your own safety make sure you have a plan to get off your vessel during any storm.

Maintain an inventory of items both left on board and removed, especially high value items, and ensure you have photographic evidence.

Ensure that you have submitted, and actioned, your windstorm plan otherwise your policy may be declared null and void.

Some questions to consider asking include: who pays for salvage especially if the boat is not a total loss? Is your mainsail covered if you lashed it to the boom? Is your Bimini frame covered if it is not stored down below? Does your policy cover equipment or belongings on the boat; and is your boat trailer included? If in doubt, speak to your agent.

Vehicle — In the event of damage to your car only fully comprehensive polices will pay out, not third party ones. You may face a hefty bill to remove your derelict vehicle.

HOUSEHOLD SUPPLIES & TOOLS

Gaffer, duct tape and WD40 will become your friend! You simply cannot have enough but you do need to keep it dry.

Tools and household items are not much good if they're stored in a part of your home that you can't get to. Store mops, buckets and brooms where you can grab them quickly in the event of flooding. Consider a waterproof plastic box with a lid for the sole use of household tools and basic items such as disposable plates/cutlery as well as a box for your initial emergency food supplies so that they are to hand. Avoid cardboard boxes as they only attract insects and other pests and are useless if they get wet.

Pre-position some essential tools, and charged batteries, in your 'safe room'.

Have a tarp or heavy duty plastic sheeting available for securing your roof or broken windows. Consider stockpiling extra plywood. Make sure you have extra nails and screws for post storm.

A rechargeable drill/screwdriver will save on time and labour and ensure you have an extra fully charged battery.

A gasoline-powered chainsaw (or a machete) is vital if you live in a remote or wooded location – many have relied on this to cut their way out of their properties – and ensure you have extra chains, fuel, spare mix oil and a chain sharpening tool or a grinder or whetstone/rock to keep your machete sharp. An 18" chainsaw is probably sufficient for most jobs. Once out, consider leaving it in your vehicle in case you have to help clear a roadway. A battery-powered saw will have sufficient charge for smaller jobs and removes the hassle of mixing gas and oil but is wholly dependent on a power source post storm.

An axe can help to remove walls or debris.

Varying lengths and thickness of rope are also useful. It can help tow a car, hold a bucket to access a cistern, or hold and secure a door in place against high winds.

It is not advisable to cut any downed power lines for fear of electrocution. Our local electricity corporation advise us that if the poles are down it is best to cut the pole and leave the power head intact rather than cut wires.

All those tins in your disaster box are no good if you cannot open them: make sure you have a non-electric can opener and waterproof matches or an ignitor to light the BBQ or gas stove.

As you may need to severely ration your water, disposable plates and cutlery can cut down on washing up or purchase a washing up container/pan as you can then recycle that water for cleaning the floors or flushing the toilet.

Purchase a whistle for each member of the household in case you need to signal for help.

Ensure your home has a portable fire extinguisher or fire blanket and a Carbon Monoxide detector if you're running a generator nearby.

Stock up on vector control items: the mosquitoes will drive you crazy. Mosquito Dunks are a great idea as mosquitoes do not travel far and tend to live close to their breeding grounds. Dunks can be placed in puddles and other standing water sources as they help kill mosquito larvae and are harmless to animals.

LIGHTS

As it's not realistic to sync our lifestyle with the sun and moon – or a bathroom break in the middle of the night - you will need options for light. And you cannot guarantee that your standby generator will survive the storm. Our brand new standby generator at work, installed two weeks previously, was damaged by the office car that was parked close by. The vehicle was inverted into the generator!

You need general lights to illuminate the living area, lights to read/play by and personal lights. Hanging torches are very useful as are headlamps for when you need both hands. You can even purchase battery operated bedside lamps. A friend ordered a lot of these after Hurricane Irma and said they make her feel a lot more normal in a power cut.

Garden solar lights outside your front door can make coming home in the dark a little less creepy and a solar light stuck in an empty wine bottle makes an effective lantern. You can also buy inflatable solar lights.

If your devices are battery operated, make sure they have fresh batteries and you have enough extra batteries in stock and have kept them dry. If the devices are not new, test them before the storm and, more importantly, before the shops close. I learnt that one the hard way! Unless you are on active alert for a storm, do not store devices with batteries in them in case they leak.

Each member of the household should have their own light source and know how to use it.

Candles and kerosene lamps are not recommended during any storm for risk of fire but are an effective option for later although I struggle to read by candle light alone. Keep any candle stubs and any drippings as you can repurpose them into new candles later. You just need new candle wicks which you can buy from any craft store.

Kerosene lamps produce a steadier light than candles and lamp oil is affordable but the smell of kerosene can become overwhelming so you would need to ensure areas are well ventilated.

POWER/ELECTRICITY

You don't truly appreciate it until you don't have it. Mains power was not restored to my colleague on the North Side of Tortola for over 182 days. Our friends in Puerto Rico's more remote areas were still without power over nine months later. One popular sports bar became a vital hub for those in the community as they could charge up their phones, use the bathrooms and get a hot meal.

I cannot believe I am saying we were lucky in the BVI to have experienced so many years of constant power cuts – and continue to do so - but at least we were used to going without power. As a result, many households and businesses already had generators or other sources of backup power. For most people, however, this is not a normal occurrence and the loss of power will bring their lifestyle to a grinding halt.

Have some form of backup power solution as electricity in some areas could be off for months even if this is just the ability to keep electronics charged. Options range from inverters, permanent standby generators, portable generators, solar power or handheld portable devices. There is something for every budget.

Gas Generators — Smaller portable generators are good for powering a few essential appliances in your home such as the refrigerator and emergency lights while standby (or fixed) generators can power most of your house.

You should decide what you want or need to keep running when the power goes out as this will determine the overall wattage of the generator you will need. A simple method of calculating wattage is to multiply the appliance's voltage by its amps. For example: mini refrigerator is 120V and 2 amps = 240 watts.

To run at optimal levels, any generator should run at 75% overall capacity to allow for a surge in power when starting most appliances.

Space is also a factor: a fixed generator will require substantial space outside on your property. Subject to wind direction, any generator should be kept at least 20 feet away from your house and any doors or windows to avoid carbon monoxide poisoning. Consider having a carbon monoxide detector in your property as a safety precaution and don't forget to have some spare parts securely stored away including extra fuel, spare oil, oil and fuel filters, coolant and fan belts.

As it is not recommended to run any generator during a cyclone – and never run one indoors - you should try to weather-proof it as water intake could short out the unit. Consider blocking the air vents with plywood and ratchet straps can help to secure a generator outdoors.

Regardless of whether your generator is under a maintenance/service contract learn the basics: your engineers may not be able to respond post storm.

Both portable and fixed generators require a constant source of fuel such as gasoline or diesel and portable generators will need frequent refuelling, running only for a few hours at a time. The larger standby generators can be hardwired by a licenced electrician to come on automatically requiring minimal work by the homeowner. Securely store your spare fuel which may need to be inside your home in a closed room.

If you hardwire the generator, do ensure that the electrician has not run the generator through the electric meter to avoid expensive surprises when the grid comes back on. I kid you not, this happened in the BVI. And you need to know how to turn the generator to manual as you will need to turn it off during the storm.

Don't run a portable generator to a wall outlet in your home in the hope of powering the home's wiring system as this could kill a utility worker working on the lines. Use an external extension lead and do not plug household appliances directly into the generator.

Gas and diesel are not your only options. Propane powered generators are affordable but they are known to be fuel thirsty and perhaps not the most cost effective option if you anticipate loss of power for a considerable time.

Solar Generators — Solar generators are typically quieter than a traditional gas generator, you can expect a good day's worth of power before needing to recharge and you can normally plug in devices directly.

However, they are not designed to power an entire home – just a few essential appliances - and they can be heavy, difficult to manoeuvre and possibly not the cheapest or most environmentally friendly option due to their batteries.

Portable Devices — Portable devices can range from solar or USB power banks, inverters, vehicle power stations, hand crank chargers or DC-AC converters. You won't power major items in your home but you will at least be able to provide yourself some light and charge any electronics. Your vehicle may even have a built-in USB adapter to charge small electronics or you can purchase a car cigarette lighter adapter to charge your phone.

If you intend on using your vehicle as a power source to charge phones or other batteries, be safe about access, alert to the risk of carbon monoxide emissions, and be aware that you will deplete the battery.

VEHICLES

When counting down to a major storm, fill up your vehicle with fuel. Power is needed to pump the fuel which might be out after the storm, and the fuel itself may be rationed initially.

Know how far (or how long) your car will go on a full tank of fuel. Consider purchasing spare fuel cans but do not store petrol inside your car or home for extended periods. It may also be possible to siphon petrol from your vehicle to fill a generator but use a pump designed for this purpose.

Make sure the vehicle is in good working condition and check the tread and air pressure on all of your tyres, including the spare. Store spare oil, wiper blades, fuses. Consider disconnecting your car battery to protect a power source post storm. My vehicle's electrics were fried, but the new battery could have been saved with hindsight.

Roads will likely be damaged with a high volume of debris and potholes to contend with. With garages likely damaged, finding an air compressor may be a challenge and a portable air compressor to keep your tyres at optimal levels is a good purchase. I have two! A tyre puncture kit is essential. The liquid 'fix a flat' is great in an emergency: two cans will take a flat tyre to safe to drive and you do not lose the tyre. Have spares as you will likely be sharing with others who have nothing.

Clear plastic sheeting or clear shower curtains/liners makes great temporary windows or windscreens but don't forget the gorilla or duct tape.

A friend advised that if you have more than one entrance/exit to your property, and more than one car, consider pre-positioning a vehicle at each end as you may then have an opportunity of getting at least one vehicle out. If you think your road will be blocked post disaster, park in a different location entirely and walk to it once it is safe to move around. Do try to park vehicles well away from trees and from water sources or flood prone areas. My friend moved his car five times before he was finally happy with the location. It is impossible to know the best place. Friends thought theirs were in a relatively sheltered spot and whilst the vehicles did not blow away, they were definitely shunted forward several feet. Thankfully they had parked one of them far enough back from the end wall that the car wasn't rammed into it, and the wind had blown in the back windows. The other vehicle suffered the same fate but additionally had a neighbour's rear brake light from two cars behind delivered into her dashboard!

You will need a very good memory in the dark and when it rains. That puddle could be the mother of all potholes!

HYGIENE & SANITATION

If there is no power you will likely have no access to running water, your dishwasher, or even a septic/sewerage system.

You will want to protect your valuable supply of stored water and keep your home clean and an effective way is to recycle any greywater for general cleaning, laundry or flushing toilets. Greywater refers to waste bath, sink and washing water which can be reused. 'Greywater' will never be safe enough to drink.

A washing up pan/bowl is useful for washing the dishes or hands in order to reuse the water, and similarly a plastic bowl can be used for bathing in the worst case scenario.

A bucket with plastic liners is an effective temporary toilet if you are not able to use the ones inside your property for whatever reason.

Store your toilet paper in plastic bags to keep it dry. Nothing will ruin your day faster than finding wet toilet paper!

MEDICAL NEEDS

Try to refill prescriptions in advance so that you have sufficient to last at least a month, possibly two, and in particular any essential medications for chronic diseases including insulin and asthma inhalers. My elderly neighbour ran out of her blood pressure medicine and the emergency clinics were overwhelmed in the first few weeks. Check with your doctor concerning the storage of any prescriptions. Insulin ideally should be kept in the fridge, but it can be kept at room temperature for up to two days if you can keep it as cool as possible. Keep all medicines dry in a sealable bag and store in a spare cooler.

Purchase or replenish your first aid kits and consider leaving one in your vehicle. Include dust masks, non-prescription drugs including antibiotic cream, pain relievers, anti-diarrhoea medicine, antacids, laxatives, activated charcoal (you can use on pets if they get into a toxic substance) and antihistamine - the wasps were bad after the hurricane. Hydrogen peroxide is not just limited for first aid, it is also an effective disinfectant cleaner.

If you suffer from any allergy, ensure you have the appropriate medication including a spare EpiPen if severe.

If you have battery operated medical equipment, have a backup battery charged and ready to go.

Lacerations and broken bones are the most common injuries and getting to the hospital or an available clinic may be difficult due to impassable roads or damaged vehicles. It is advisable to take a basic first aid course or look online for instructional videos. You can make, for example, a basic splint from a newspaper or a pillow case. At the minimum, keep a first aid book with your contingency supplies.

CASH

Get as much cash out as you can afford. Bank services are likely going to be disrupted and your credit/debit and ATM cards won't work without power. Available shops and garages/petrol stations may insist on cash only. Outgoing flights may also require payment in cash if you are trying to leave the area.

COOKING OPTIONS

You need to prepare healthy meals as there's a limit to how much Vienna sausage you can eat out of a can!

Most households in the Caribbean have a BBQ and so long as you have a lot of aluminium foil there's actually not much you can't cook on it – basically, it's just upside down grilling! Before the storm check you are full on propane and consider having a spare tank or stock up on charcoal and lighter fluid. Don't forget the matches and store both charcoal and matches in waterproof containers…

So long as the ignitor isn't electric, a gas cooker/stove should work just fine with matches. As for a gas BBQ check your propane level.

If you have limited power after the storm – for example at the office because of a generator – do consider the merits of a slow cooker (Crockpot). A friend had power at work (not at home) and there would be a line of slow cookers at her office cooking meals for the staff to take home.

COMMUNICATIONS

Hurricane Irma damaged most of the BVI's mobile telephone network, the telephone poles for landlines, antennas, sirens and repeaters. Those with VHF radios were able to communicate with each other but most of the community were cut off in the first few days, including officials, resulting in 'runners' being used to relay messages.

WhatsApp became invaluable for many of us in the lead up to Irma. We had different work groups but a friend created a group chat on 4 September 2017 so that, no matter where we were in the BVI, we could check up on each other and share information. As I re-read the discussion almost two years later, individual resilience and humour still shines through despite unbelievable concern about what was ahead. The chat stopped the morning of 6 September 2017 when the network went down.

Communications Plan — Have a communications plan in advance with loved ones outside of the affected area and with affected friends or family if you won't be together. It could take some time to get word of your safety and media reports and images of a catastrophic scene can cause untold anxiety on loved ones. Everyone needs to be on the same page about what to do, and who to contact, in a worst case scenario.

The only reliable method of communication to the outside world for us in the first few days after Irma was a satellite phone. But they need airtime to work through a subscription, pre-paid SIM or pre-purchased minutes which can be expensive. If you purchase the airtime independently of the phone supplier you may need to arrange to have the phone unlocked. Satellite phones also depend on satellite coverage so you should check that your area is included. If you intend to rely on a satellite phone to communicate with others also holding satellite devices, arrange check in times pre

> ## RUMOURS WILL BE RAMPANT. USE YOUR BEST JUDGEMENT AND DON'T PANIC.

and post storm to ensure everyone is connected at the same time.

Your plan should include a method of leaving messages for each other post storm if you share the same house. Close friends used to leave each other post it notes on the fridge to let each other know where they were to avoid worry and concern.

Know your alternative methods of communication: text messages and phone calls may still work even if data is down.

Sign up for text alerts and warnings from your local and regional authorities and buy a battery-operated AM/FM radio to receive updates and emergency broadcasts from local radio stations in the event you lose power. Some telecommunications companies offer location based services such as traffic reports and evacuation routes.

Make sure you have paid your bill. Our local telecommunications companies either reinstated disconnected customers or extended payment deadlines but you cannot guarantee it. Make sure that your phone has been charged and holds a full battery. You may wish to consider having a spare, fully charged, battery as back-up. Have a plan for keeping your phone dry. If you have a landline, purchase a non-cordless telephone and keep this in your disaster box. If you lose power, and landlines are working, your phone should then work.

Ensure your grab bag contains call credit, charged back-up phone, portable chargers, and power banks, Save important contacts including family members to this spare phone as well as writing them down.

Emergency Communications — Satellite devices vary in capability and price. They range from top of the line (and budget) VSAT terminals, BGANs, handheld satellite phones, GPS communicators to satellite Wi-Fi hotspots. There is a high likelihood that telecommunications will be knocked out and it is recommended that you explore one of these options, splitting the cost with neighbours if necessary.

VHF radios were also heavily utilised at the marina where I lived not just for general communication but to also make security alerts within the residential community at the height of the looting. They were the fastest way to share new weather bulletins: we were facing down Hurricane Jose within days of Irma and no ability to track given the lack of internet or phone. I ended up checking daily with our Disaster Adviser based in Miami who would give me an update via satellite phone and I would relay to the marina community via VHF. I will forever remember the cheers that resounded around the marina when we realised Jose would miss us.

Another useful device is a handheld UHF radio. They are cost effective, don't need a licence and line of sight normally quite significant.

CLOTHING

After the hurricane had passed and especially during clean-up efforts I do wish I had had a pair of Wellington Boots or safety boots to protect my feet, and a friend who lives on Virgin Gorda discovered that her new motorcycle boots became a sought after commodity!

A good friend was staying at a house which was destroyed during Hurricane Irma and they had very little time to get to safety. Like many people she lives in flip-flops and she lost these during the storm and had to navigate broken glass and nails barefoot. When I approached her to suggest one piece of advice to others in a similar situation she gave me two: "wear proper shoes" and "have a bag attached to you which includes dry clothes". She had thought ahead and packed a change of clothes in her car but lost the car keys in the ensuing panic. She said they were wet for days. I, and many others, heeded her advice and slept in closed toe shoes during the night passing of Hurricane Maria two short weeks later.

Many people I have spoken to relay the same complaint – wet feet. Try to have a spare pair of dry shoes/boots and plenty of clean, dry socks.

Walking around in wet clothing causes chafing.

Trust me on this: at least two weeks' worth of clean underwear. Such a morale boost when you cannot easily do laundry.

You will need appropriate clothing for clean-up. Quick-dry long sleeved shirts and long trousers are recommended and have the added bonus of protecting you from mosquito bites. You will need multiple pairs of work gloves for clearing debris and glass. Sturdy boots/shoes will protect your feet and don't forget spare socks.

Do consider leaving a complete change of clothing for each member of the household in a backup location. Your vehicle is one option (but make sure the keys are on your person) or another secure location preferably on high ground.

Additional advice has now been shared from The Bahamas: a helmet (motorcycle, bicycle or construction hard hat) can help to avoid head injuries from flying debris.

Wear a bright coloured shirt in case you need to be spotted and denim jeans can double up as either a life preserver or cut up and repurposed. If you regrettably find yourself having to defend yourself against the elements, foul weather gear will not only keep you dry, it will help to protect you: rain drops travelling at high speeds can feel like needles all over your body.

Don't forget your pets' paws – broken glass and nails can hurt them too.

PAPERWORK

Storing your important papers in a waterproof box or on an external hard drive is sensible but if the box or hard drive ends up down the hill because the contents of your house were sucked out it won't do you much good. Keep the originals in a waterproof and secure container and scan documents and store them electronically in the cloud.

Write down any important numbers or medical information and keep securely – if you lose your phone, or cannot charge it, you lose all your contacts.

Consider sharing bank account numbers, credit card numbers, login details, and copies of vital documents (such as passports, insurance, wills, deeds, birth certificates) with a relative who will not be impacted by the storm. My colleague shared her family's passport and visa details with her sister in advance of Hurricane Irma who was able to make all travel bookings on their behalf when the family decided to leave after the storm but had no internet ability to do it directly.

ENTERTAINMENT

BVI's Cable TV network was knocked out during the hurricane and remains out almost two years later.

No internet means no streaming of live TV, movies online, social media or email.

Plan ahead for your entertainment for during the storm, especially if you have young children, and potentially afterwards. Have family board games, packs of cards, books and any other games that don't require batteries or power. Consider downloading movies to your device assuming you have a means of recharging it.

It is important to maintain as much normalcy in the family routine as possible. There might not be any TV to watch but we saw many families and couples reconnect over games of monopoly or chats over candlelight.

FOREIGN NATIONALS & PASSPORTS

Check the validity on your passport before the official cyclone season begins and renew in plenty of time. If you need to go to another country, it is highly unlikely you will be allowed to enter without a valid travel document and replacements can take some time. You may also need a visa.

Foreign nationals, whether resident or visiting, should consider registering with their nearest Embassy or Consulate if this service is offered and should ensure they have the relevant contact details. Unaffected family members should also have these details.

Keep your passport on your person during the storm at all times.

If tourists or foreign residents are advised to leave before the storm impact, GO! You could become a burden on fast depleting food and water supplies and other valuable resources. The main airport may also be closed to routine air traffic after the storm to allow entry for military, aid and other humanitarian traffic.

SUMMARY

Learn your local weather resources. Identify your 'safe room' or alternative location in advance and clear exterior areas of any projectiles. Pre-position some supplies in your 'safe room'. Reach out to any neighbours, friends or family members that are alone. Ensure any travel documents are valid. Prepare on at least two to three weeks of food and water for each member of the household including your animals. Ensure you have sufficient insurance, medications, cash, fuel, household tools and cleaning supplies and power sources. Identify communications options in the event of phone failures. Factor in time to prepare and secure your boat.

REFLECTION FROM GUS JASPERT
GOVERNOR OF THE VIRGIN ISLANDS (UK)

I started as Governor of the British Virgin Islands (BVI) two weeks before hurricane Irma hit. On arrival, the Territory was still overcoming severe flooding from the worst rains in 50 years, but attention quickly turned to preparing for what was coming across the Atlantic. My family hadn't even started unpacking all our bags before preparations commenced. As a family you hope your first shopping trip to stock the cupboards in the new home isn't to get cans of tinned food and emergency water supplies!

As Governor I was thrown into national preparations with the Premier, Ministers, the Department of Disaster Management and critical agencies and voluntary organisations. Governmental level preparations are vital. And leadership at a national level is critical in times of crisis. But the experience of Irma and Maria highlighted how vital individual's preparations are and their own leadership with their families and communities. Claire Hunter's excellent book provides a practical guide for just that. Some of these are lessons that have been learnt the hard way - from the experience of being hit by the biggest recorded hurricane and living through two Cat 5s in two weeks. No-one is above preparations and hurricanes are a great leveller. Government House may be a grand building, but, like most of the population, my family and I with some colleagues did just what most people did and sheltered in the smallest safest place of the building – the bathroom. To this day many BVI conversations still touch at some point on people's 'bathroom stories'!

Amidst the tips, anecdotes and some light hearted reflections, there is a serious message and guide in this book. The need to think through in advance how to keep yourself, your family, your business, your home, your boat and what you care about as safe as possible.

Tragically we lost lives in the hurricanes of 2017. Without preparation we undoubtedly would have lost more. Hurricanes are measured by the Saffir Simpson scale and it's worth remembering that beyond the technical analysis of wind speeds - the impact descriptions are sobering. Even a Cat 1 can have 'very dangerous winds that will produce some damage'; and for a Cat 5 'extremely catastrophic damage will occur'. In the face of impending danger and catastrophe this book gives a good run through of how to reduce the impact it could have on you; and also how to bounce back quickly afterwards. Sadly, this book also looks like it will be more necessary in the future as climatic events become more frequent and potentially more devastating as we witness the impact of climate change.

This book also contains a sense of the spirit of the BVI. The hurricanes of 2017 were extremely catastrophic and images of the devastated buildings, destroyed infrastructure, obliterated nature take a while to overcome mentally. But, what won't ever fade in memories is the spirit of the people of the BVI – the strength and resilience as communities came together to help each other and to make sure that whilst knocked down, we weren't knocked out; and in the face of such challenge to still smile and look to the future.

PART 2

RESPOND

There were five of us huddled on my kitchen floor during Hurricane Irma plus three dogs and the abandoned marina cat. Our animals were kept on leash and the leash attached to a human.

Although the condo was elevated and we had previously confirmed we were not in an evacuation zone, we were worried about potential storm surge and we discussed Plan 'B' (and 'C'). Everybody had a job including my friend's 11 year old son which was to grab the cat in her bag if we had to move.

TALL RUBBER BOOTS ARE ESPECIALLY GOOD IF THE AREA IS FLOODED!

The storm surge came to the lip of the balcony on one side and near the front door on the other but thankfully receded preventing us from having to leave. My landlord told me months later that he was afraid the front door would blow on the back of the eye and he was right. Despite a waterfall through the ceiling and through the now door-less doors we were very lucky.

We found out that the North side of Tortola was rammed with 40ft sea surges; my colleague on the other side of the marina had been flooded through from sea surge; and my best friend, were it not for the heroic efforts of her husband, was almost sucked out of a window in her hillside home by a tornado. When Hurricane Maria came through two weeks' later, she hid in a cupboard.

DURING THE STORM

Go to your 'safe room' away from flood prone areas and windows and doors – an interior hallway/corridor or windowless bathroom is best – and remain there with a backup light source and extra batteries if the device is not solar.

Do not lean against windows or doors in an attempt to hold them as this only puts yourself in danger.

Do not go outside and do not go down to the water to watch the storm as you could be caught in large waves, storm surges, flood waters or flying debris.

If you find yourself trapped by flooding, go to the highest level of the building if you can but avoid your attic unless you have direct access to the roof.

Falling barometric pressure doesn't remove your roof but the wind will. Keep your (shuttered) windows closed to prevent wind from entering your home. Some claim opening windows on opposite sides of the home to counter pressure is a myth, but many did this during Irma and swear it alleviated the pressure on doors and windows. My friend told me over a year later that the wooden floorboards upstairs in my condo were buckling under the pressure which subsided once the front door and balcony doors blew out.

Keep identification on your person at all times, including your passport. You may not have time to take your 'grab bag'.

Wear proper shoes and consider hanging a whistle around your neck in case you need to signal for help. Avoid alcohol during the storm. My biggest regret of all.

Expect the unexpected. Changes in pressure will make your ears pop and the water level in your toilet bowl will slosh around. Remain alert and ready for just about anything to happen.

THE EYE

Remember that the eye will bring some calm but winds will rapidly come back up to tropical force strength and you may have insufficient time to get back to safety. How long do you have? Depending on the size and speed of the storm, the eye can last anywhere from a few minutes to a few hours. Wait for the all clear and do not attempt to leave unless your safety is of paramount concern or you need to make quick emergency repairs.

CHILDREN

Your response will only be as good as you have planned. Now is not the time to search for games, grab bags or look for snacks or water. Hunker down in your 'safe room' and remain as calm as you can. Explain to your children, in an age appropriate way, about what is happening and distract as much as humanly possible through games, movies and singing.

ANIMALS

Keep your pets with you in your 'safe room' and preferably on leash or in a kennel or bag. If your domestic animals are afraid, do not underestimate minimising their senses. You can cover their bag/kennel with a towel or blanket and reassure them.

Back in the day when my dog Lucy would go out on a boat, she would brace herself when out in choppy waters and big waves. During Irma, my friend's dog, Maci (she was emaciated when she was rescued), would physically brace herself

NEVER UNDERESTIMATE PET PREPARATION

SWIM FINS ARE NOT PROPER FOOTWEAR!

whenever there would be a significant drop in barometric pressure or when the relentless winds would try to penetrate the building. This became a vital forewarning for us humans although I vividly remember the extreme discomfort in my ears whenever the pressure would drop.

GENERATOR

As tempting as it is to run your generator when the mains power has been turned off, don't. You are wasting precious fuel and risk shorting out your generator if it intakes any water.

COMMUNICATIONS

Keep your phone on charge but if the mobile and data networks go down, turn your phone off to conserve power. Limit non-emergency mobile phone calls to avoid overloading the network and use internet based messaging or text messaging instead. If using a landline, opt for a corded telephone and avoid unnecessarily long calls. If you have a VHF radio, keep this on your person, fully charged.

FLOODING

If you find yourself trapped by flooding, go to the highest level of the building if you can but avoid your attic unless you have direct access to the roof.

ALL CLEAR

The method of the ALL CLEAR will be determined by your local authorities and you should find out what your local arrangements are in advance. In the BVI we had no communication and the sirens were destroyed. We relied on previous experience to determine when Irma was finally moving through and it was safe to move around.

SUMMARY

Stay calm and track the storm if possible. Keep identification on your person at all times, including your passport. Go to an interior room and remain there until the 'all clear'. Do not run your generator during the storm or go outside during the eye unless safety of paramount concern or you need to make quick emergency repairs. Wear proper shoes and have a whistle.

REFLECTION FROM JOHN S DUNCAN OBE
GOVERNOR OF THE VIRGIN ISLANDS (UK) 2014 - 2017

2017 was an especially challenging extreme weather year for the BVI.
Even before hurricanes Irma and Maria struck the Islands we had experienced some of the worst flash floods in living memory. Natural disasters are not only shocking for families and individuals, they cause massive disruption to government and local services. Perhaps the most serious challenge facing those in positions of leadership is to discover what exactly is happening, how bad is the situation, where should I focus my effort? Communication, both to learn the answers to these questions, to be able to direct help to where it is most needed and to reassure the public that someone is going to respond, is absolutely essential. But in most crisis situations normal means of communication eg. telephone will probably be badly damaged or simply overloaded. Alternative 'back up' systems need to be varied. Both 'old' technology - VHF, SW radio and the latest technology FaceTime, WhatsApp messaging, all played a part during the crisis BVI faced and overcame in 2017.

Claire Hunter was a key player in the crisis response team in Tortola. She played an essential role in laying the groundwork to allow local leadership, public services and NGO's to respond when the crisis came, but equally important was her role in preparing the public for the inevitable communication 'black zone' so that they had taken the key steps needed to protect themselves. She has written an admirable practical insider's guide based on her personal and professional experience of both the 2017 and previous crises. I am sure it will serve as a key reference for others in the future.

REFLECTION FROM BOYD MCCLEARY
GOVERNOR OF THE VIRGIN ISLANDS (UK) 2010 - 2014

I experienced my first tropical cyclone, Hurricane Earl, 10 days after arriving in Tortola and this was swiftly followed by the terrible rains brought by Tropical Storm Otto. Earl reported in the local press as being my "baptism of fire". And so it was.

The things that struck me most about these events was:

- the devastating power of the wind and rain
- the unevenness and unpredictability of the impact; Earl brought much stronger winds, but Otto did much more damage
- the professionalism of the DDM under the outstanding leadership of Sharleen Dabreo
- the vulnerability of the weakest members of the community
- the amazing resilience of the Virgin Islanders; no sooner had the storms passed than people were out assessing the damage and starting to clear up

As my colleagues have stressed, the key advice I have to offer is for everyone, businesses and households, to prepare as effectively as possible. Nothing says this better than the DDM moto: "It is better to prepare and prevent rather than repair and repent".

I add my congratulations to Claire Hunter, one of the longest-serving members of the Governor's team, for having put together this guide.

PART 3

RECOVER

Be prepared for shock, disbelief, complete exhaustion, panic, chaos and the need to think on your feet. You will become expert at 'jury rigging' repairs. Adrenalin will help you keep going but you cannot sustain at that pace indefinitely.

Be neighbourly. Reach out to those in your community that may need some extra assistance whether they are elderly or simply alone.

Many in the community have joked by saying that Hurricane Irma was an alcoholic. Despite the considerable damage to buildings, liquor cabinets were left intact. One of my colleagues told me about her sister's home: she had a solid granite kitchen island with the only item left on it a bottle of champagne. Irma smashed the counter into pieces but the bottle of champagne had been knocked to the floor and sat there unbroken. Another colleague relayed that when they ventured out of their 'safe room' (which had a wine cellar!) to discover the rest of the property had been wrecked, her husband went in search of any supplies that may have survived. He found the booze and the Kitchen Aid was in the pool.

DINNER FOR ONE

WATER

Use bottled water, or water you have properly prepared and stored for emergency use, for drinking, cooking, brushing your teeth and for washing the dishes.

If you have insufficient drinking water, boil it for at least 1-3 minutes, longer if possible, to kill spores. If the water is cloudy before boiling, let it settle and filter through a clean cloth or coffee filter. Let the water cool and store it in a covered clean container.

If you cannot boil water, use fresh unscented liquid bleach (1/4 teaspoon bleach (5.25-8.25%) per gallon when water is first stored). If the water remains cloudy, double the amount of bleach. Stir and let it stand for at least 60 minutes. The water should have a slight chlorine smell. If it doesn't, repeat and let it stand for a further 15 minutes.

Other water sources that could be treated include melting ice cubes, draining your hot water tank or pipes or other flowing water such as from a river or lake. Do not use water with any floating material or water that has a dark colour or strange smell!

CHILDREN

I referenced my friend earlier whose young daughter went through Hurricane Irma oblivious in good quality headphones. They were not in their own home and had taken shelter elsewhere with another friend plus various dogs that were being looked after. This property was fairly remote and another friend, concerned about their wellbeing, had gone to find them after the hurricane passed. Despite the significant damage to the property and sheer volume of debris and downed trees, they were thankfully okay. During the long walk home in the heat, my friend told me that her daughter's only clothes were sparkly wellington boots, leggings and a tutu. She said despite the devastation and shock throughout the community, seeing this blonde three-year old child walking in her tutu brought many smiles.

My friend also recommended making sure you are well stocked on drinks and snacks no matter where you go. Many known snack stops may have been destroyed or are closed, young children don't understand why, and she said it is simply not worth the meltdown over a candy bar.

Let your kids be kids, safety permitting. Encourage as much play and interaction with their peers as possible but don't let them play in or near flood waters.

If you have access to the news, experts recommend monitoring and limiting media exposure to children. The coverage, and images, may cause information overload. Try to return to a normal routine as much as possible, even if you are not in your own home, and involve your children in any clean-up efforts.

If your children appear in shock and are cold, give them some warmer clothes to wear. Signs of stress and anxiety can include nightmares, disrupted sleep, stomach upsets and changes in behaviour. Encourage your children to ask questions and help others. Most families will recover given enough time, self-care and a return to some normalcy, but seek professional help if reactions don't improve, or get worse.

Many children were sent away after Hurricane Irma, some away for months.

Schools were closed, homes were damaged, power and telecommunications were patchy, and many families faced financial hardship. The parents I spoke to who sent their children away felt this was a double-edged sword: they knew they were doing the best thing for their children but felt guilty. For the kids that remained, they suffered the loss of their daily routine as well as many of their friends.

If your children do need to leave encourage a new school environment even for a short time. This is part of a normal routine.

If you have relocated to a public shelter, keep your children safe. Encourage a 'buddy system' and advise your children to avoid isolated areas. Accompany very young children to the bathrooms and shower rooms and ensure sleeping arrangements are appropriate.

ANIMALS

Do not use the storm and damaged properties, gates or fences as an excuse to let your animals roam: roaming animals creates a danger, not to mention a huge annoyance, to other animals and those in the community. Secure your animals another way, including tying them up.

Clear your property of debris and any mosquito breeding sources. Use Mosquito Dunks in puddles or other standing water that cannot be removed. Mosquitos do not travel far from their breeding ground and the bite of one mosquito infected with the heartworm larvae will give your dog heartworm disease if they are not being treated with preventative medicine.

There was truly an international response to the plight of all animals in the BVI post hurricane. One heart-warming story involved the saving of a rescued German Shepherd who severed an artery on some debris. Basic first aid was administered to the dog but with no means to contact anyone, and the dog's normal vet away, the prognosis was not good. In desperation, the owner and friends found their way to the emergency room at the hospital and met a visiting doctor from Colorado on a break in-between shifts. On seeing the dog in the back of the car this young doctor agreed to help and jumped in the vehicle to the veterinary clinic which was functioning as a pet evacuation centre run by the Puerto Rican Humane Society. The doctor and the Vet Tech then went to work to save the dog. Another great human then volunteered his dog as a blood donor. The visiting doctor said it was one of the most rewarding things she had done in her career, the dog is fine and apparently really enjoys going to the vet for check-ups. His owner believes it is down to seeing the Vet Tech and to show his appreciation!

Domestic — "Vomit of Woe" … our 'sister' veterinary clinic in the US Virgin Islands relayed the story of their clients needing assistance post hurricane. It was basic human nature after such a traumatic event but each and every one of those customers needed to share their personal story of survival before they were able to communicate what they needed from the clinic. Everyone was listened to and I understand shots of rum offered …

My colleague had her cats in the office after Irma as her home had been destroyed. She discovered disposable roasting pans/tins make great cat litter trays. The vets tell me that you do not need cat litter: soil is just as effective as are paper towels as the cats don't really care … if you want to treat them, tear the paper up.

If drinking water is being conserved, wet (tinned) food is a good idea and will help hydrate them.

Walk dogs on a lead until they become re-oriented to their home. Familiar scents and landmarks may be altered and animals could easily become disoriented and lost. Debris, downed power lines, and reptiles brought in with high water can also pose a danger for any animal.

Days are long and tiring after a catastrophic storm. If you need animal care during the day and know others with similar difficulties – and the animals get along of course – consider setting up a dog or cat crèche during the day. You can identify a 'point' person and house and all animals left there for the day.

My friend lost one of her cats during Hurricane Irma and searched for months afterwards. It was a miracle but she eventually found him, but along the way found a cat she had previously fostered and rehomed. The new owner had had the opportunity of leaving the Territory and left the cat who was now forced to learn to survive alone. Don't abandon them: your animals are dependent on you for food, shelter and safety.

If you are evacuating your pet, attach an accurate photo with its name to the top of the crate in case of escape.

I am told that if you run out of dog poop bags, and you happen to have latex gloves in your first aid kit, these do the job just as well …

Livestock & Horses — My friend, the Vet Tech, who has a farm and retired/ rescued racehorses, said horse owners around the world are selfless: she does not know what she would have done without their generosity. Her farm was devastated and she relayed the story that her horses re-appeared one by one at the ridge of her hill looking down at the farm. She felt they were surveying the damage knowing life was going to be very tough: her horses did not eat for ten days before a kind benefactor donated a 40 foot container full of hay.

If feed is in short supply, or gone, bulk existing supplies with plain rice (whole wheat is best). Animals need fuel for energy.

Chlorine tablets will help keep water sources clean from bacteria. If you have lost your water source for livestock or horses, you may need to get creative.

RE-ENTERING YOUR HOME

One resident, away for the storm, couldn't comprehend that her house was completely destroyed when her son-in-law managed to check in. She was a little obsessed that the church cash she had in a drawer was still there and, on watching video footage, identified her Silk Cut cigarettes lying in debris. Thanks Mum: yes, we're still alive!

Enter your home with caution. Wear protective clothing and gloves when clearing up debris and proper shoes to prevent cutting feet on sharp debris. It is best not to work alone.

Open windows and doors to ventilate and dry your home. Use a flashlight to inspect damage. Never use candles and other open flames indoors until your utilities have been checked for leaks.

Do not touch electrical equipment if it is wet or if you are standing in water. Turn on your main breaker only if safe to do so. I learnt this the hard way when I attempted to turn on my waterlogged ceiling fan.

Protect your property by boarding up broken windows and tarping your roof to help deter additional weather damage or looting. I begged a spare padlock and hardware from a neighbour to block off my gate as I no longer had a front door, and a large piece of debris was found and nailed in its place to attempt to keep looters out.

Watch out for wasps! We have a particularly nasty species in the BVI called a Jack Spaniard and they were everywhere after the hurricane.

Be aware that flood water brings with it the risk of waterborne bacterial contaminations and uninvited guests such as snakes and rats.

Damaged rental properties, especially in a small community, will create 'supply and demand' and unless protective legislation is in place you can expect to see rents go up and tenants evicted.

Beware of unlicensed contractors who may give you a discount but leave you an even bigger mess to clean up.

HYGIENE & SANITATION

The risk of disease after a natural disaster is a realistic possibility when there may be a total loss of power and clean running water. There are three major sources of disease: water, food and vector. Practice good hygiene habits to reduce any risk and take great care if you are buying cooked items from pop up food vendors. The following may help.

General hygiene & bathing — Consider having a separate bowl/basin in your bathroom for washing your hands and/or cleaning yourself. This water can then be used for flushing toilets.

Hand sanitiser or wet wipes will cut back on the amount of water being used.

A solar marine or camp shower can hang outside to heat up and brought inside for a shower. Friends used a watering can as an interim shower.

Laundry — This will not be an immediate priority unless you are trying to salvage flood damaged items and, if items are dry, you will wear every last piece. Your easiest option is to hand wash and line dry. The water can again be reused for general cleaning. Friends had a morning routine and used two baby baths: one to wash/soak and one to rinse. Their clothing choices were always quick dry/tech materials that dried quickly.

Toilets — The mantra in the house I was sharing was "if it's yellow let it mellow; if it's brown flush it down!"

As the marina's sewerage system was damaged during the hurricane, residents were asked to initially avoid using the toilets and we had a temporary toilet outside using a bucket. Once the military had arrived in Territory (and were using our marina as a base) one of my housemates remarked over dinner that the helicopter pilot had waved at her. Another housemate exclaimed "well, he didn't bloody wave at me when I was sat on the bucket!"

A bucket with plastic liners is an effective temporary toilet if you are not able to use the ones inside your property for whatever reason. Consider having two: one for urine and one for faeces. Mix one cup of liquid bleach with two quarts of water and (carefully!) pour into the lined bucket. Add more of this mixture after each use and change the bag when it becomes half full. If you cannot bury it in a deep trench (at least 2-3 feet) away from water sources, tie the bag carefully and place into a separate lined container with a tight fitting lid.

Consider keeping basic supplies in the bucket for user convenience such as toilet paper, wet wipes, extra liners and the disinfectant supplies.

If your waste empties into a septic tank you have to avoid certain chemicals or at least use them sparingly.

If the sewerage lines are open, but you have no running water, all you need to do is dump a bucket of greywater into the toilet bowl with sufficient force to create a strong flush rather than using the toilet lever to access the tank supply in the normal way. Have some towels ready to mop up any mess and don't forget to raise the toilet seat.

Washing up — If there's no power and limited water, your dishwasher will not work.

Make sure plates are scraped of food before it has a chance to dry. Boil the water on the stove if you can and fill a small container in your sink. We had a separate pot just for rinse water and both the washing up and rinse water was reused for other cleaning jobs as well as potentially flushing the toilets. If you plug the sink, you cannot reuse the water. Disposable cloths to clean work surfaces will cut down on water and dirty towels.

Solid Waste — The BVI's Department of Waste Management reported a year after Irma that they had collected half a million cubic yards of debris on the main island of Tortola alone. Bearing in mind that an average year generated 30,000 tons of residential waste, the volume was unprecedented and caused enormous pressure on local authorities. An alternative rubbish site to the main incinerator (damaged during the hurricane) had been established but material going in was not sorted and residents dumped other household waste resulting in a massive amount of combustible material. Local NGOs and private recyclers tried, and continue to try, to promote and manage recycling initiatives. Not unsurprisingly, fires broke out necessitating a cross agency response and continue to smoulder almost two years on.

Our Chief Fire Officer recommended that if your main solid waste facility is overwhelmed, individuals should make efforts within their own communities to sort their debris and general waste until local authorities are in a position to step in to collect and manage it.

Waste streams should be divided into three key areas: anything that is easily burnt (wood, paper, plastic); anything that is glass or resembles glass (such as ceramics); and anything metallic. Food scraps, not meat or dairy, can be buried to allow it to decompose naturally – 8 inches of soil should keep most rodents out. Other food waste should be securely stored in a rubbish bin with a lid.

If you cannot bury human waste in a deep trench (at least 2-3 feet) away from water sources, tie the bag carefully and place into a separate lined container with a tight fitting lid until local facilities allow proper disposal.

COOKING & FOOD

Guacamole is what my friend and colleague found on her balcony as she ventured outside at the all clear. She had been eying up almost ripe avocados growing next to her apartment building in the days before Irma but did not get an opportunity to pick any. Irma did that for her and plastered them all over her balcony! If you're lucky to have, or live near, fruit trees try to salvage any fruit that high winds may have brought down. They will be a nutritious and tasty treat you will appreciate.

You do not need to cook three main meals a day especially if you are surviving on contingency stocks. Plan on one hot meal per day with smaller, more casual meals or snacks for the rest of the day. One pot dinners are the best.

Before you start opening new supplies, cook anything perishable in your freezer or refrigerator first.

My ex-boyfriend is obsessed with canned vegetables to this day and cannot walk past a supermarket without buying a can or three. He was not prepared before the storm and turned up at my home with a five-gallon bottle of water, some bottles of wine and cigarettes. He finally moved back into his own apartment six weeks later.

PEST & VECTOR CONTROL

I have had Dengue and Zika and don't recommend it.

Flies come out from nowhere – so do the mosquitoes – all bringing with them the threat of spreading disease. Between unprecedented winds that damaged or destroyed buildings creating high volumes of debris, torrential rain and storm surge, ideal breeding conditions are created for mosquitoes and other pests. You should take the normal precautions against being bitten. Citronella candles, incense and repellent sprays all help. You may need a mosquito net to sleep under or screens at the door or windows and you may appreciate a battery operated or solar powered fan to keep them at bay. I acquired some hanging sticky fly paper which successfully trapped dozens of annoying flies – but keep it out of the reach of domestic animals.

Mosquitoes need standing water to breed but don't travel great distances. You can help reduce mosquito populations by draining or covering any potential water source including storm debris, rain gutters, buckets, flower pots, tyres and dog bowls. Mosquito dunks help for areas, such as puddles, that you cannot remove. Flooded swimming pools can also be a problem. For pet owners, be aware that the bite of one mosquito infected with the heartworm larvae will give your dog heartworm disease if they are not being treated with preventative medicine.

Rodents have lost their homes too and will be in search of food, water and shelter and they are coming to you! Remove food sources, water and any items that potentially provide shelter. Rats are aggressive, dangerous and will find a route in. If you can identify any openings do your best to block it. My friend met a rat trying to enter her home via the bathroom shower drain … Whac-A-Mole anyone?!

Get rid of your rubbish on a regular basis. It is sensible to sort your rubbish at the outset: use a secure container with a lid for any perishable food items and store dry waste and any debris elsewhere. This is particularly important if rubbish collection services are suspended. If you see signs of rodent activity, thoroughly clean and disinfect the area (and utensils) to reduce exposure to any diseases.

SECURITY

Have photo identification on your person. Security operations may include

checkpoints and local authorities may implement a curfew. Apart from mitigating security concerns, this curfew also gives emergency responders the opportunity to clear the roads of debris without the interference of normal traffic.

Our marina was heavily looted and the 'coconut telegraph' active about imminent attacks forcing lockdown for residents and limited access to non-residents. A very large catamaran had blocked the main entrance to the marina which became a useful blockade and all available people took turns in doing security. It wasn't just our marina that was looted, damaged homes across the islands were equally preyed upon. You may need to consider a neighbourhood watch in your area.

I was recently reminded about a couple of stories about looting which happened in the days following Hurricane Irma at the marina where I live. The first morning after the hurricane, an acquaintance appeared at my balcony carrying as many items of clothing as he could from one of the marina shops begging me for safe storage. His friend's store was being raided and he was trying to save as much inventory as he could. A looter was also seen running away carrying a large salad bowl and another dodging large boats on the pedestrian dock as he attempted to run with a stolen dock cart over his head. I had hoped he might trip and fall in. Another acquaintance went to see what was left in a small supermarket in Road Town and chuckled later at the evening community meeting that local looters clearly didn't realise that Perrier was water. Apparently there were cases of bottles untouched at the back of the store. The moral of this story? Crisis does not always bring out the best in people.

GENERATORS & BACK UP POWER

The primary power source after Hurricane Irma were generators.

Minimise run time and maximise your load. Long term damage can occur to your generator if you overload it and it will likely fail if you run it at 100% for any length of time due to surge loadings when the motors start. Ideally run your generator at a 50-75% load.

Do not run any generator or BBQ indoors: the carbon monoxide will kill you in minutes.

If you share your generator with neighbours do consider establishing set times each day that it will be used so that they can plan around that. And have a plan with those sharing to cover the costs of fuel as well as taking it in turns to get it.

COMMUNICATIONS

Whilst the Territory's communications were knocked down, we weren't knocked out thanks in large part to the BVI's Amateur Radio network.

BVI's 'Ham' operators were active before Irma's impact and were able to speak to fellow operators in Dominica who kept the information coming about what to expect. Despite extensive damage to the telecommunications network and the failure of regular communications channels, the Amateur Radio League's repeater survived the hurricane and they were able to broadcast messages on special frequencies to individuals and critical agencies, such as CDEMA, outside of the Territory.

In any disaster, Hams are vital but it appears to be a dying skill and one challenge the BVI faced was not having sufficient radio operators on the sister islands. They are now making a big push to increase numbers and encourage people from all walks of life to get involved.

I have a good friend who is the IT Manager at the marina where I live. Despite the local challenges, he partnered with Digicel who had a working tower in Road Town and created a network by bouncing the signal into the marina via two properties on the mountain, having set them up with antennas, solar power and generators. He deserves a medal for his brilliance in establishing an internet hotspot. In true reflection of his sense of humour the limited network was named 'F U Irma'. Unfortunately, many abused the limited network by downloading massive files or streaming videos of the hurricane footage and certain websites required disabling.

Even if mobile networks are up, try to limit non-emergency phone calls to avoid overloading the network. It is best to use SMS/text messages or messages over the internet such as WhatsApp and/or social media to communicate instead or use a landline. If cellular coverage is intermittent, texting has the added benefit of being able to send whenever you enter a working network.

Conserve your mobile phone battery by reducing the brightness of your screen, turning off Bluetooth and/or Wi-Fi unless necessary and closing unnecessary applications. If you have no other power source, remember you can charge your phone in your vehicle but only if safe to do so.

Word of mouth also became critical. People walking past a friend's house were able to share names of others they had seen as well as relaying the wellbeing of my friends. It took days to hear news of some people but this network was vital and very reassuring. My colleague and I used this approach when we worked through the missing lists and could account for people we had personally seen.

If you have access to a satellite phone and multiple calls to make in the same country, consider appointing a family member or friend there to make the calls domestically. It is hard receiving a call from a complete stranger when the conversation begins with "it's about your daughter in the BVI ..." but when the first words are "she is safe" especially not having had any word for days, the news is priceless.

Emergency Communications — Satellite phones are handy but understand that airtime is expensive. If your satellite device enables the internet, a smartphone or tablet will tether less bandwidth than a laptop. Remember to turn off tethering when not in use (even for a short time). Turn off any automatic updates on your device and limit any streaming of videos or sending attachments to avoid unnecessary charges.

If you have access to a satellite phone and multiple calls to make in the same country, consider appointing a family member or friend there to make the calls domestically. It is hard receiving a call from a complete stranger when the conversation begins with "it's about your daughter in the BVI ..." but when the first words are "she is safe" especially not having had any word for days, the news is priceless.

The sister island of Jost Van Dyke was cut off from any information. One of the many stories I heard was that of our volunteer Marine Search and Rescue team (VISAR) who discovered this. They set up a scheduled VHF radio check every day by positioning volunteers on the main island of Tortola with line of sight to a specific point on Jost, and a member of the community could check in and receive updates. Think outside of the box.

Social Media — I will blame 'brain fog' immediately after Hurricane Irma, but despite my experience on satellite phones managed to convince myself they wouldn't work at night! This of course is not true. As a result, I only made contact with my mother the following day to reassure her I was okay and to ask her to post on my Facebook page my safekeeping as well as tagging the names of the people I was with.

Another lifeline was a Facebook page created pre hurricane by BVI expatriates. The page was established so that friends and family outside of the BVI had a way of searching for accurate information. Use every method of communication available to you, and learn about ones specific to your situation.

There was an official list to report your safekeeping in the BVI with the Red Cross, but with no advance information, and no reliable communication after the storm, many residents had no way of knowing about it. Multiple versions were then created on a personal basis but only caused more confusion. Facebook, for example, does offer a crisis response tool called 'Safety Check'. If this has been activated, users have an opportunity to mark themselves safe and potentially check on the status of other affected people.

GETTING AROUND

I had no transport after the storm as my vehicle was destroyed. And in the first few days there was a concern about safety with the widespread looting and the Prison breach. If I needed to get to work to get information, I was taken in on a dirt bike with an ex-serviceman friend for protection. I remember sitting on a wall outside Government House, waiting on my ride back to the marina, being surprised by the sheer number of vehicles out on the roads sightseeing and thinking about the waste of petrol!

Do not walk, swim or drive through flood waters. Six inches of fast moving water can knock you off your feet and one foot of moving water can sweep your vehicle away.

Wait until an area is declared safe before entering and watch for closed roads. If you come across a barricade or flooded road, turn around.

Standing water may be electrically charged from power lines and metal fences may have been "energised" by fallen wires.

Avoid downed or dangling utility wires, especially when cutting or clearing fallen trees. Our local electricity corporation advise that if the poles are down it is best to cut the pole and leave the power head intact rather than cut wires.

I know I go on about it, but a further reminder to wear proper shoes to protect your feet from sharp debris, broken glass and nails. Debris is under water too.

VEHICLES

My old Pathfinder was 'Irma-ed' (who needs windows anyway). I had also been looking after my neighbour's car as he was away. One of my hurricane buddies had gone outside briefly during the eye to get a general sense of the damage (yes, we shouted at him), and when he came back in he said "...don't worry about any dents in Roger's car." When I questioned this, he replied "because there's a tree in the

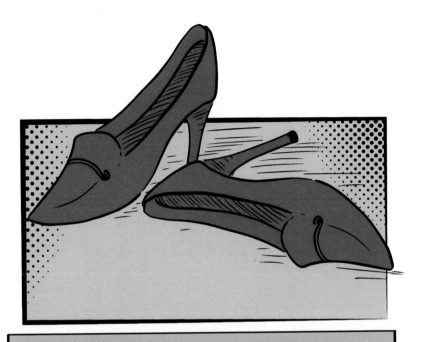

LOOKING GOOD DOESN'T APPLY AFTER A CAT 5 STORM. GET SOME BOOTS.

windscreen!"

My then boyfriend's Jeep Wrangler became a convertible after something sliced the roof off. We became very creative in finding ways to cover the vehicle – and went through a massive amount of Gorilla tape - but as it turns out a vehicle without a roof will always be a wet ride when it rains!

Not many vehicles escaped damage and the community became quite creative in making them driveable. Clear plastic and tape became quite the precious commodities and a friend recently reminded me about seeing vanity mirrors being used as side mirrors! One friend managed to fix up her broken car windows with Bimini plastic and Velcro so that the windows could be opened, which are still in place today. Hug a sailmaker! I can remember driving with a friend to what was left of his house in a small pickup with no windscreen. It started to rain and he put the windscreen wipers on. Routine and habits are hard to break.

BOATS

You made it through the storm and you now have time to go check your boat(s). A friend in the industry has recommended the following.

Check below and bilges. Find the source of any water ingress and plug it with whatever is handy.

TRAINERS ARE NOT IDEAL, BUT "OKAY"

Check your hull very thoroughly above the water line – many boats that seemed fine have sunk because of a small hole above the water line that was not originally seen.

Do a thorough engine and battery visual check before powering anything up with batteries, or turning on the engine.

Seal any damaged deck with tape and/or silicone to keep water out.

Let your salvage operator know you are insured and who your insurance company is.

If your engine(s) were submerged during the storm do not remove them from the water until you have mechanic(s) ready to start working on them. As soon as the engines come out of the water start working on getting them running. Let your mechanic know that you turned off electricity to the motor before the storm. If you have a marine gear, have it separated from the engine so it does not get seized in place.

If your boat is repairable and the insurance is going to pay for the repair, do not settle on the value of the repair until the repairs are done or your contractor has put a hard quote to fix your boat.

Remember that storage costs are usually part of the repair bill. You should hire a surveyor of your choice that represents your interests solely.

SALVAGING WET ITEMS

I know too many people who were diagnosed with Pneumonia post disaster and I am sure this was as a direct result of mould.

Move out saturated items including mattresses, upholstery and area rugs. If you intend to claim on insurance you may need to store these items until they have been inspected. As I was claiming on insurance and it took some time to get an adjustor scheduled, I had very smelly rugs on my balcony accumulating more mould. It was impossible with all the rain that followed Irma to dry them out properly.

Clean out remaining debris and mud and disinfect the floors and remaining objects and wear protective gloves. If you have foam padding under any carpets, remove and dry outside.

If your phone gets wet remove the battery immediately, gently wipe off as much moisture as possible, wrap in a dry paper towel and immerse in a bag of uncooked rice for 24+ hours. I've successfully used this tip after my phone fell in a puddle.

One of my colleagues was not in her own home for the hurricane and I, and another colleague, had to salvage her ground floor. She had stored her contingency supplies in cardboard boxes which had disintegrated in the storm water. We threw away any item that was not in waterproof packaging and liberated the rum! If this happens to you, all is not lost. Remove labels from wet cans, disinfect the can and use a permanent marker to note the contents. If cookware has come into contact with contaminated water you can wash them in a bleach and water solution to disinfect,

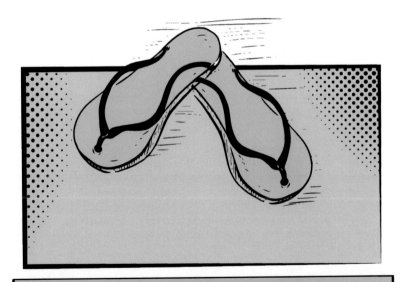

SANDALS LEAVE MOST OF YOUR FEET EXPOSED TO SHARP OBJECTS. AVOID.

but for safety, best to throw away anything made of wood or plastic.

If you can, and once power has been addressed, freeze wet documents including books and passports unless they have been exposed to any sewerage. They can remain in the freezer indefinitely and will eventually dry there. I did this for a close friend whose passport I found when helping clear what was left of his business premises.

Lay wet paintings face up and allow to air dry. Do not remove from their frames. However, do remove any wet framed prints and drawings from their frames and air dry. If you cannot remove any wet matting then air dry with the mat.

Wet photos can be air dried glossy side up. If stuck together, place in a tray of water and allow them to come apart naturally or freeze them.

Air dry wet clothes immediately and shake out any debris. If you cannot wash items immediately, store the now dry clothes in a clean plastic bag.

SWIMMING POOL

Many swimming pools in the BVI turned completely black after Hurricane Irma which is not uncommon when a storm is as bad as she was.

Please remember that your pool professional, if you use one, may also be affected by the storm and unavailable or uncontactable for a few days. Our friendly pool expert was trapped at home for four days after part of the hillside collapsed and her business premises were also destroyed.

Depending on the condition of your pool post storm, you may have no choice but to dump all the water from the pool due to contamination but this could oversaturate the ground around it and create further problems. Apply common sense as it could become a costly exercise treating the contaminated water that ultimately you have to throw away.

If you have determined that your pool is saveable, remove any debris from the pool and check electrical equipment before restarting the system if safe to do so. Adjust the chemical levels in the water as needed, add a dose of shock, and walk away.

BUSINESSES/OFFICES

Once safe to do so, a designated member of staff should visit the business premises to assess the damage and compile an action plan highlighting immediate security or health and safety concerns. Consideration should be given to activating the business continuity plan and all staff notified.

Once back at work, managers should monitor staff performance and look for signs of any exhaustion/stress/health risks and make sure those staff are relieved of their duties for necessary rest and support.

INSURANCE CLAIMS

Be patient: your insurance company will likely be overwhelmed and if they provide instructions for filing claims, including deadlines for quotes, or steps to take to protect your assets from further damage, make sure to follow them.

Marine — One of my favourite insurance stories was a gentleman calling from

overseas to demand an update on the status of his boat and to know why a marine surveyor had not yet inspected his vessel.

I've had some fun at this customer's expense to make the point that people overseas had no concept of what the reality on the ground was.

If your boat is damaged, or sunk, this is when you need to know the terms of your insurance policy. Will they pay the salvage operator? Will they pay for haul out and storage and repairs? What is your deductible? Start communicating and coordinating with your insurance company, salvage operator, boat yard, marine surveyor and insurance adjusters.

You should get a copy of the survey report from the insurance marine survey and/or insurance adjusters. This is your right.

If your boat is so badly damaged that it is not worth the cost of repairing it, you should receive two offers from your insurance company. One is payment to you and releasing all rights to the boat. Insured value minus the deductible. The other is an offer from the insurance company minus the 'residual' value of the boat and you retain the ownership of the vessel. Both of these cases are relative to the term in your policy: Constructive Total Loss.

Depending on the insurance coverage you have taken out, physical depreciation is not normally factored into determining the value of the lost or damaged vessel, but sails, canvas, trailers and some machinery might.

If your boat is repairable and the insurance is going to pay for the repair, do not settle on the value of the repair until the repairs are done or your contractor has put a hard quote to fix your boat.

Household — Create a list of any damaged contents against your original inventory and document any damage with photographs. Contact your insurance company as soon as you can for further advice or details of any time limit to claim and do not throw away any damaged items until an adjuster has said you can.

If you haven't yet read the small print on your policy do it now to avoid surprises later: your deductible may be larger than you think and your claim smaller than you realise.

Keep accurate records of your expenses and save bills/receipts from temporary repairs.

Do not make permanent repairs until an insurance adjuster has reviewed the damage and given the go-ahead. When the adjuster arrives, ask for official identification and contact information.

Vehicle — To make a claim on your vehicle insurance for damages caused by a natural disaster you will need to hold comprehensive coverage. The cost of repair following serious flood damage to your vehicle's engine and electrical systems will likely exceed the vehicle's value. If you can push your insurance company to 'total' your vehicle, do it.

DISPLACEMENT & MULTIPLE OCCUPANCY

Although the marina where I live was badly damaged, the residential properties had held up quite well. The complex became a safe haven for so many displaced in the community, and in the first few days resembled a refugee camp as weary individuals sought shelter. Most condos became multiple living. A few hotel rooms had barely survived but were also used.

I will be forever grateful to dog walking friends who became lifelong friends after taking us in after the hurricane. It must have been quite a challenge having seven people and five dogs living in a two bedroom condo disrupting their normal routine but they did it with great kindness and great humour.

If you are living multiple occupancy, do not underestimate the morale boost of sitting down together for a hot meal each evening. The kind friends that took us in insisted on this normalcy and I am very grateful.

Establish 'house rules' and have a plan for asking for, and contributing to, household expenses including generator costs. Being out of pocket all the time is the fastest way to kill a friendship.

At some point, and subject to your personal means, you may need to consider whether leaving the area is the best decision especially if you could become a burden on fast depleting supplies and accommodation.

EMOTIONAL WELLBEING

It wasn't just the humans that were traumatised by the storm. My young dog is still frightened at noises off the water or a simple thunderstorm. I think she remembers when the front door blew in and the sound of the wind and driving rain. I remember feeling like I was in a 'fog' for days with a total inability to concentrate and even now a random image of the hurricane will bring a wave of emotion – gone a few seconds later.

Regardless of whether you are living multiple occupancy, or lucky to remain in your own home, with your family or new hurricane family, understand that days are busy, long and stressful with everybody working in very difficult conditions. I cannot stress enough the importance of a hot meal together every evening with a debrief of everyone's day.

A friend told me that their new 'normal' became BBQs with friends that were displaced or just struggling to cope – they had a lot of people descend on them for reassurance, food and alcohol and not necessarily in that order!

Help where you can: us Brits are known for making a cup of tea in stressful situations and my 90 year old neighbour made enormous volumes of tea every day for those individuals clearing debris and salvaging the boats. Others just let strangers talk, cooked extra food, ran errands for people, used personal equipment and tools to help clear roads, or threw their chainsaw over their shoulder and just got to it!

The term Post Traumatic Stress Disorder (PTSD) will be used a lot to describe how people are feeling. My best advice to anyone affected is to seek support from friends and family, rest, talk about your experience, get back into a normal routine, and to treat yourself with kindness. If you do suffer serious long-term effects seek professional help. There is no normal or abnormal reaction as everyone is different. You might experience many emotions: tearfulness, sadness, anger or even shame or regret – or nothing at all. It is all normal. Laughter helps relieve tension and take a day off whenever you can.

FOREIGN NATIONALS

Any national wishing to leave the country after the storm must be mindful that subject to damage, and the local security situation, the main airport may be closed initially to commercial air traffic to allow entry for military, aid and other humanitarian traffic. Other methods of entry and exit may be similarly affected such as the ports. Irma saw much resourcefulness in the BVI community for those wanting to leave including private boats to Puerto Rico. Non US nationals or non US Permanent Residents were not permitted in the immediate aftermath to enter the US Virgin Islands by any method as they were managing their own security situation.

Depending on the severity of the storm, some governments may send a plane for their nationals wishing to leave after the storm. Depending on the distance this may not be directly to the home country. Luggage is normally restricted to what you can carry; no pets are generally allowed; and it's unlikely to be free.

EMERGENCY RELIEF

Donations can become additional debris. Help from outside will likely be needed in addition to regional and international partners, and all efforts appreciated.

However, well-intentioned but unsolicited support can fast become a burden on already stretched local resources if it's not coordinated properly and items close to expiry dates (such as medicines) can exacerbate rubbish disposal problems if solid waste services have been suspended. Wait until the local authorities have completed their immediate needs assessment before organising relief goods, and remember that monetary donations are sometimes better as they go to longer term recovery efforts.

If you are organising any fundraiser or crowdfunding, be transparent about its purpose, how you will get any funds raised to the beneficiary, and keep proper financial records.

Beware of criminals taking advantage of tragic events and donate only to organisations or charities that you know and trust.

Civilian aircraft wishing to assist relief efforts should contact local Civil Aviation and disaster authorities before they fly to ensure they have the appropriate approval.

That being said, I don't believe sufficient thanks has been extended to our friends in Puerto Rico in the immediate aftermath of Hurricane Irma. The 'Puerto Rican Navy' as we affectionately call them diligently persevered through the red tape to come across with essential supplies and filled volumes of private boats at their own expense to help the Territory. Multiple containers of relief, and planes, then followed. Shortly afterwards, category 4 Hurricane Maria hit Puerto Rico and we were not in a position to help them. Thank you.

Aid — To avoid extra work for local disaster authorities, coordinate in advance to ensure essential items are obtained and consigned properly. Many donations of relief items were sent to the BVI with no local coordination and unsolicited containers clogged up the port causing major delays in clearance.

Don't send items in unmarked boxes. Everything should have a proper packing list, bill of lading, and shipping pallets can be shrink wrapped. Think ahead to how the items might need to be stored.

If you donate generators, consider despatching with a spare parts package including oil, oil filters as well as fuel filters as these items may not be available locally for some time. The same goes for chainsaws.

Bear in mind any cultural differences in food: what might be an obvious staple in your pantry might not be familiar to others and could risk sitting uneaten on the shelves of the local Food Bank.

If you donate hay or feed for livestock and horses, consider donating the shipping container too. The recipient may not have alternative or secure storage options.

If you donate veterinary items (eg vaccines), coordinate with your local veterinary practice or Humane Society International who are experienced in disaster response for animals. To help with the storage of such items, you can also consider donating small battery or solar powered refrigerators, cool boxes or containers with dry ice.

Volunteers — If you intend to volunteer, either as part of a larger non-government organisation (NGO) or individually, make sure you have coordinated with the local authorities. Many jurisdictions will expect all volunteers to be registered with the host government and working in support of the national relief and recovery effort. You risk demoralising national relief workers by duplicating their work. Many volunteers arrived in Territory Post Irma as tourists and only declared themselves as NGOs once on the ground. This created a burden on local authorities as many required additional support not previously anticipated and volunteer efforts duplicated government led work.

If you are a registered or licenced professional, approach local authorities in advance as they will have provisions for emergency registrations.

Volunteers, and media outlets, should bring resilience, plans for accommodation, and their own supplies of food and water. Too many didn't and ended up begging supplies from those who had already lost everything they owned.

Fundraising — You don't need 15 people to coordinate funds. Identify one external person to pull in the funds and coordinate any shipments and have one person at the other end managing the receipt.

There were many reputable fundraising events that companies did for their staff or others in the community to help offset the costs of replacing personal items or buying generators or even food. Many people had lost everything they owned. I vividly remember seeing individuals in the main town pushing a shopping cart containing the sum of what was left of their homes and feeling overwhelmingly sad. But fundraising requests have a timeframe before people grow weary of giving money and most grants available from the government or other charitable organisations will simply not be there 6 months later.

If you live in an area at risk from tropical cyclones, try to have insurance or some savings tucked away.

Environment — It wasn't until days after Irma that it occurred to me that I couldn't remember when I had last seen the birds or the delightful hummingbird that sat on a branch off my balcony every day. Even the little geckos had disappeared and there wasn't an ant in sight. It was about six months following the storm that butterflies started appearing again and even frogs seemed to have become over-populated. The point here is that the natural environment has an incredible way of readjusting itself after an intense and devastating storm.

Depending on the intensity of the cyclone, wildlife and their natural habitats can be drastically altered within a matter of hours. Some wildlife may get a sense of falling barometric pressure or other changes and bug out (no pun intended) early while others may suffer from wind displacement and end up hundreds of miles away from their natural habitat and species of their own.

Food supplies for wildlife may become scarce or even non-existent unless you are able to provide birdseed that you have stockpiled before the storm or find fruit that may be close to being spoiled. For some nectar-eating birds and bats, sugar water (one part white sugar, four parts water and placed in a small container in an area easily accessible by them) is going to be most welcomed.

The once lush hillsides can be completely stripped of foliage and, in the case of Irma, even the bark was stripped off trees. This extreme change of scenery seems to have a psychological effect on people as it was a constant reminder of what had happened. When vegetation begins to return over the following months after an extreme storm, morale also seems to increase as a positive sign of recovery.

However, be aware of any non-native species that can quickly spread, smother or out-compete the native species. Native species are more resilient to future storms which may help provide coastal protection or better stabilise hillsides. Removing or cutting down any 'dead looking' mangroves immediately following a storm may result in a complete loss of the habitat as mangroves take longer to recover, particularly red mangroves. Waiting at least three months before even attempting to cut or prune mangroves is critical and even then, removal will increase vulnerability to coastal erosion and loss of habitat for birds in an already stressed environment.

Another change in the environment are dramatic changes in natural stormwater courses because of the volume of storm debris or fallen trees blocking these channels. This may require you to notify the local authorities as soon as possible as these changes can result in new channels being formed. Water will always find its way downhill and dramatic changes in stormwater flow could potentially result in landslides and flooding within areas that had never flooded previously.

Marine ecosystems will also be severely impacted, particularly the shallower, nearshore coral reefs and seagrass beds from storm surge and sediment transport. Even vegetation ripped from the shoreline may end up on the reef along with storm debris. Underwater debris might provide new habitat for fish and invertebrates but it should be removed to ensure the ecosystem isn't further damaged in the event of another high wave energy event.

If you discover large volumes of red algae in the water after a storm, don't panic. Large quantities of red algae were found immediately after the 2017 hurricanes and was caused by the catastrophic tides. If it's still there eighteen months later, however, then re-address water quality issues.

SUMMARY

Be prepared for shock and exhaustion. Safety first: re-enter homes and businesses with caution and avoid flooded roads. Practice good hygiene habits and reduce mosquito and rat breeding areas by separating your rubbish and clearing your yard of any debris. Keep your children safe especially when in public shelters. Cook any perishable food items first before opening new supplies and when using a generator minimise run time and maximise your load. Secure your animals: the storm is not an excuse. Volunteers and Relief Agencies should try to identify what is required before making arrangements to send or travel. Remove any underwater debris, protect mangroves and help feed wildlife that have lost food sources.

REFLECTION FROM FRANK J SAVAGE CMG, LVO, OBE
GOVERNOR OF THE VIRGIN ISLANDS (UK) 1998 - 2002

I am pleased to make a small contribution to Claire Hunter's excellent, and compelling, book "Surviving Irmageddon." The lessons of which, if followed will undoubtedly save lives particularly among excessively vulnerable members of families and within the community and in particular among the very young and elderly.

In my ten years living in the Caribbean, firstly as the Governor of Montserrat (which included nearly 30 months managing the response to an erupting volcano which eventually left the capital, Plymouth, buried under 80' of volcanic debris- now a modern day Pompeii) and then as the Governor of the British Virgin Islands, my wife and I went through six hurricanes of varying strengths and destructive power. They were all terrifying but with advance planning, we came through them safely and even enriched by the experience. After retirement I was engaged by the Foreign and Commonwealth Office as its Disaster Management Adviser to ensure that disaster mitigation measures were in place in all the UK Overseas Territories and was called upon to assist Governors in Cayman and the Turks and Caicos Islands with the aftermath of massive hurricanes in 2004 and 2008. One lesson I have learned in all this time, is that personal and family preparedness is of vital importance and you ignore it at your peril. Governments have a duty to prepare their country and community for a wide range of disasters but the responsibility for your own safety, and that of your family's safety, rests with each of us individually.

We were fortunate in both Montserrat and the BVI that we had outstanding local disaster preparedness professionals who conducted comprehensive educational and awareness programmes to prepare their communities, as a consequence of which there were mercifully many fewer fatalities and serious injuries than in many of the surrounding Caribbean islands.

It is indisputable that good personal and family preparations before a hurricane can and do reduce by a very wide margin the risk of injury and death. Claire, with whom I worked closely for several years in Tortola, has produced this excellent book based upon her experiences during Hurricane Irma and which if followed will save lives. Possibly yours, or a member of your family! Please read it and adapt it for your own personal family circumstances.

REFLECTION FROM DAVID PEAREY
GOVERNOR OF THE VIRGIN ISLANDS (UK) 2006 - 2010

I am so impressed by the effort and thought that has gone into preparing this guide which I trust will now become essential reading for anyone, anywhere threatened by tropical cyclones. A decade ago during my own time in the British Virgin Islands I was fortunate not to experience anything more powerful than a Category 3 hurricane, though the human and financial consequences of that particular event were quite serious enough. Ten years on we are only too well aware that with global warming comes an ever increasing risk of severe weather events.

Never certain what surprises each hurricane season might spring, the lessons I learned from my first days in the BVI were threefold: (a) the crucial role of the community in ensuring that the most vulnerable amongst us were never overlooked, (b) the essential work of the disaster management experts in mitigating the risks at every stage of a crisis - the BVI's team was, and still is, amongst the very best and has over the years saved many lives, and (c) the absolute need for individual awareness and preparedness.

This guide addresses the latter need with a host of practical advice for the layman. I congratulate the author, Claire Hunter, for producing a most readable and informative manual. Her years of working in the disaster management field are all now encapsulated in these pages. I urge you all (every year!) to follow its advice closely and so help keep you, your family and friends safe.

ACKNOWLEDGEMENTS

This book would not have been possible without the immense support from the following people and companies:

Sharleen Dabreo Lettsome for encouraging me to "think bigger" than a 'shoulda coulda woulda' list and to turn the idea into a book, followed by eighteen months of relentless support to see it through.

Corporate and private sponsorship from Caribbean Insurers Ltd; Brickell Travel Management, Miami; CreatiVertical; Nanny Cay Resort & Marina, Harmony Yoga BVI; Maria & Ben Mays; Rebecca Clark; Kath Adams: for recognising the importance of the message to encourage people to be better prepared and for making the design and production of this book possible.

The sharing of expertise: Henry Leonnig (Horizon Yacht Charters); Dr Shannon Gore (Coastal Management Consulting & Association of Reef Keepers); Dr Sarah Weston (Canines Cats & Critters); Alison Knights Bramble (Countryside Adventures at Diamond Estate Farm); Tim & Belinda Dabbs (Marine Maintenance Services (BVI) Ltd); Dr Ronald Georges (Chief Executive Officer, BVI Health Services Authority); Zebalon McLean (Chief Fire Officer, Virgin Islands Fire & Rescue Service); Rayma Blackett (Crystal Pools Ltd); Dion Stoutt (STO Enterprise Ltd); Thor Downing (Roger Downing & Partner Co Ltd); Chris Conway (Civil & Structural Engineering Ltd); Jasen Penn (BVI Amateur Radio League); Digicel (Dominica) Ltd; Tamsin Rand.

Online resources including: Centers for Disease Control and Prevention (CDC); Federal Emergency Management Agency (FEMA); Florida Department of Health; Library of Congress; Northeast Document Conservation Center; Ready.gov; The Horse Fund/International Fund for Horses; The Humane Society of the United States; The National Child Traumatic Stress Network (NCTSN); The National Oceanic and Atmospheric Administration (NOAA); Washington State Department of Health; World Health Organisation (WHO);

My family, friends and colleagues for endless edits, suggestions and support. Karen Slater, Maria Mays, Julia Campbell, Cherryl Fahie, Sadiqua Chinnery, Cayley Smit – for their resilience through Hurricane Irma and continued good humour. You make going to work in difficult conditions fun!

Kwasi Wesselhoft, Mary Storie and Jennifer Maltarp – thank you.

Finally, Sharon and Hugo Maltarp for taking me (and my dogs) in after Hurricane Irma, keeping me safe and introducing me to authentic Poutine!

CHECKLISTS

The following pages contain lists for preparation, surviving and recovering from a tropical cyclone. Print them off in case your devices aren't operable during or after a storm.

PRE-SEASON

☐ Check clip/straps on roof for excessive corrosion

☐ Check and test all shutters

☐ Check all window/door screws are in place and are not corroded

☐ Check locks/hinges/latches on exterior doors and replace if required

☐ Check cistern overflow screens are easily removed

☐ Agree your family storm and communications plan

☐ Review insurance policies

☐ Confirm your boat plan

☐ Assemble contingency supplies (household & pets)

ALERT

☐ Identify 'safe room' or alternative property

☐ Identify satphone check in times

☐ Check insurance policies

☐ Contact vet to check on pet paperwork & check pet medicine supplies

☐ Secure personal documents including wills, insurance policies, passports, medical information etc

- ☐ Check contingency supplies

- ☐ Assemble pet disaster kit (if required)

- ☐ Restock personal medical/first aid kit

- ☐ Do not pack refrigerator/freezer with perishable foods

- ☐ Monitor weather

- ☐ Charge & test emergency communications, mobile phones, IT, lights etc

- ☐ Obtain cash

- ☐ Secure your boat

- ☐ Clear & disconnect/block gutters & trim trees/shrubbery

WATCH

- ☐ Refuel vehicle & check oil, water & tyres

- ☐ Identify where you will park vehicle

- ☐ If relevant, know how to isolate generator & refuel

- ☐ Know how to isolate utilities & assemble relevant tools

- ☐ Check drinking water supplies & store clean water for hygiene/cooking

- ☐ Refill prescriptions

- ☐ Clear outside areas & secure items

- ☐ Protect doors/windows

- ☐ Leave swimming pool filled & super-chlorinated

- ☐ Turn refrigerator/freezer to highest setting

- ☐ Pack a 'grab bag'

☐ Take photos of valuables for insurance claims

☐ If flood prone, roll up carpets & store valuables in plastic & move possessions to higher floor

☐ Make alternative arrangements for your pets (if you cannot take them with you)

WARNING

☐ Complete preparedness activities

☐ Bolt doors & secure windows

☐ Close all interior doors & prepare your 'safe room'

☐ Bring pets indoors

☐ Switch off generator (if relevant) – do not use indoors

☐ Shut off utilities at main switches

☐ Shut off valve to propane tanks, disconnect and securely store

☐ Continue charging solar devices

☐ Keep household items in a plastic box & cleaning supplies to hand for easy access

☐ Write down important numbers

☐ Let others know where you are going if you leave & leave enough time

☐ Pre-position at least 2 days' worth of non-perishable food, water, first aid kit, baby supplies, grab bags and essential tools in your 'safe room'

☐ Unplug electronic appliances and devices

☐ Weatherproof any exterior breaker panels (and generator if relevant)

☐ Remove cistern overflow screens

☐ Go into the storm well fed and hydrate (with water, not rum!)

☐ Disconnect car battery

DURING

☐ Sit tight & ensure personal safety. Stay away from doors/windows

☐ Track storm with VHF or battery operated radio

☐ Wear a whistle

☐ Do not go outside during the eye

☐ Do not run your generator

☐ Expect the unexpected

☐ Wear proper shoes or boots

☐ Keep pets close (on lead or in travel bag)

☐ Avoid candles and kerosene lamps

☐ Keep your ID/passport, emergency credit card & spare cash on your person

POST EVENT

☐ Check in whereabouts using pre-arranged times or locations

☐ Keep monitoring weather

☐ Wait until an area is safe before entering & watch for closed roads

☐ Standing water might be electrically charged

☐ Avoid hanging utility wires

- ☐ Beware of snakes/insects driven to higher ground

- ☐ Do not use tap water until officials confirm safe - disinfect

- ☐ Use phone for emergencies only - send text messages if telecommunications down/patchy

- ☐ Have photo ID

- ☐ Avoid driving/sightseeing

- ☐ Do not dump in ghuts/ditches

- ☐ Keep pets on lead - do not let them roam

- ☐ Start your vehicle immediately

- ☐ Open doors/windows to ventilate

- ☐ Wear proper shoes or boots

- ☐ Enter home with caution

- ☐ Salvage wet or damaged items & prepare insurance loss claim

- ☐ Practice good hygiene habits - separate rubbish streams and keep secure

- ☐ Generators - minimise run time & maximise load

- ☐ Cook anything perishable first

- ☐ Reduce mosquito breeding opportunities

- ☐ Rest & return to a normal routine

- ☐ Be neighborly

RECOMMENDED SUPPLIES & TOOLS

The following pages contain lists for supplies and tools which will aid you in preparation, surviving and recovering from a tropical cyclone. Print them off in case your devices aren't operable during or after a storm.

WATER

☐ Minimum 1/2 gallon drinking water per person per day for at least 2-3 weeks

☐ Minimum 1/2 gallon of potable water per person per day for food preparation and sanitation

☐ Fill coolers with ice

FOOD

NON-PERISHABLE PACKAGED/CANNED FOOD FOR AT LEAST 2-3 WEEKS PER PERSON INCLUDING:

☐ Ready to eat canned meat, fruit, pulses & vegetables

☐ Meal replacement drinks or MREs

☐ Canned/boxed juice

☐ Cereal

☐ Soup

☐ Peanut butter/trail mix/granola bars

☐ Instant coffee/tea

- ☐ Instant or long life milk

- ☐ Sugar

- ☐ Bread/crackers/biscuits

- ☐ Fresh fruit (apples, oranges, pears)

- ☐ Special food for infants & elderly

- ☐ Comfort Foods

BABY NEEDS

- ☐ Special foods (at least 2-3 weeks)

- ☐ Formula (at least 2-3 weeks)

- ☐ Extra diapers

- ☐ Medicines (get a copy of prescription)

- ☐ Blankets

- ☐ Diaper Rash Ointment

- ☐ Baby Wipes

- ☐ Powder

- ☐ Bottles

- ☐ Pacifier

- ☐ Favourite toy/blanket

HEALTH & FIRST AID KIT

- ☐ Adhesive bandages (assorted sizes)

- ☐ Safety pins/catches
- ☐ Sterile eye pad
- ☐ Sterile all-purpose dressings
- ☐ Butterfly closures
- ☐ Finger splint
- ☐ Adhesive tape
- ☐ Latex gloves
- ☐ Instant cold pack
- ☐ Antibiotics & ointment
- ☐ Pain reliever
- ☐ Medication
- ☐ Anti-diarrhoea medicine
- ☐ Anti-inflammatory
- ☐ Antihistamines
- ☐ Hydrocortisone
- ☐ Dust mask
- ☐ 1st Aid handbook/guide
- ☐ Prescriptions – at least 1-2 months
- ☐ Vitamins – at least 1-2 months
- ☐ Mosquito repellent
- ☐ Sunscreen

- ☐ Sunburn/aloe vera

- ☐ Tweezers

- ☐ Rehydration drinks

- ☐ Toiletries/personal hygiene/ feminine supplies

- ☐ Eyeglasses/contacts & solution

- ☐ Hydrogen Peroxide

- ☐ Blanket/survival wrap

- ☐ Hand sanitizer

- ☐ Medicine dropper

HOUSEHOLD

- ☐ Freeze water in bags/jugs or buy ice

- ☐ 2x coolers (ice & food storage)

- ☐ Torches/lanterns/flashlights + extra batteries or solar

- ☐ Solar chargers

- ☐ Battery or solar fan

- ☐ Candles/kerosene lamps

- ☐ Carbon monoxide detector (generator)

- ☐ Fire Blanket or fire extinguisher

- ☐ Battery or solar radio

- ☐ VHF or UHF radio

- ☐ Phone chargers (including one for vehicle)

- ☐ Grill + extra propane, charcoal or sterno
- ☐ Slow cooker/crockpot
- ☐ Matches or butane lighter
- ☐ Disposable plates/cutlery
- ☐ Manual can opener
- ☐ Aluminium foil
- ☐ Uncooked Rice (wet phone or for animals)
- ☐ Permanent marker
- ☐ Plastic storage containers
- ☐ Fly traps
- ☐ Mosquito traps/incense etc/Mosquito dunks (for standing water)
- ☐ Mosquito net

SANITATION

- ☐ Toilet paper, soap, baby wipes, liquid hand sanitizer
- ☐ Plastic bags
- ☐ Liquid detergent
- ☐ Unscented liquid bleach (fresh)
- ☐ Plastic garbage bags, ties (for personal sanitation uses)
- ☐ Plastic bucket with tight lid
- ☐ Disinfectant or wipes
- ☐ Plenty of absorbent towels

☐ Rubber gloves

☐ Hydrogen Peroxide

☐ Rubber gloves

☐ Bucket/mop/broom

☐ Paper towels

☐ Drain cleaning product

☐ Water purification tablets

☐ Baby baths (for washing clothes)

☐ Plastic container/tub (washing/rinsing dishes)

ELECTRONICS

☐ Phone Prepaid Minutes / Airtime

☐ Cellphone, tablet & chargers

☐ Spare SIM / spare phone / call credit

☐ Satellite phone & charger

☐ VHF radio, spare battery or UHF radio

☐ Battery operated AM/FM radio

TOOLS

☐ Basic tool kit

☐ Utility knife/Leatherman

☐ Axe

- ☐ Wrench

- ☐ Bolt cutters

- ☐ Pliers

- ☐ Battery operated drill/screwdriver

- ☐ Hammer, Nails, screws, masonry screws

- ☐ Handsaw/machete/chainsaw, extra fuel, spare mix oil, chains & chain sharpening tool, whetstone

- ☐ Rope

- ☐ Work gloves

- ☐ WD40

- ☐ Duct/gorilla tape

- ☐ External extension cords

- ☐ Shovel/rake

- ☐ Padlocks

- ☐ Ladder

- ☐ Plywood

ITEMS FOR GRAB BAG

- ☐ Change of clothing per person

- ☐ Cash

- ☐ Emergency credit card

- ☐ Emergency communications (satphone, mobile phone, call credit & radio)

- [ ] Portable charger

- [ ] Important papers (eg passport, insurance, key contact numbers, list of family doctors & serial numbers of medical devices, pet papers)

- [ ] Water & portable container

- [ ] High energy snacks

- [ ] Flashlight/Torch & extra batteries

- [ ] Utility knife

- [ ] Rope

- [ ] Work gloves

- [ ] Essential medication

- [ ] Extra house/car keys

- [ ] Photo ID

- [ ] Valuable items (eg jewellery)

- [ ] Basic toiletries (travel size)

- [ ] Sunscreen

- [ ] Mosquito Repellent

- [ ] Animal food/water/container/spare leash & muzzle

- [ ] Whistle

- [ ] Ground sheet/blanket

- [ ] Wetwipes/hand sanitiser

- [ ] Matches/lighter

- [ ] Walking shoes & socks

VEHICLE

☐ Jumper cables for vehicle

☐ Portable air compressor

☐ Tyre patch kit

☐ Portable Tyre Infation Kit in aerosol can

☐ Work gloves

☐ Glow sticks

☐ First aid kit

☐ Long sleeved shirt, trousers, closed shoes

☐ Machete or spare chain saw

☐ Tow Rope/Chain

FOR EVACUATION OR SHELTER

☐ Pillows, blankets, sleeping bags or air mattresses.

☐ Folding chairs or cots

☐ Extra clothing and shoes

☐ Personal hygiene items: toothbrush, washcloth, deodorant, etc.

☐ Food, water, ice (supply of non-perishable canned foods)

☐ Manual can opener

☐ Portable radio and batteries

☐ Flashlights and batteries

- [ ] Prescription medications in their original containers

- [ ] Books, handheld games, cards, toys, needlework

- [ ] Eye mask & ear plugs

- [ ] Cash

- [ ] Battery/solar powered fan

- [ ] Important papers but travel light (including driver's license, special medical information, insurance policies, property inventories and passport).

MISCELLANEOUS

- [ ] Back-up power source (eg inverter or generator with spare oil, oil filters, fuel filters)

- [ ] Cash (at least 2-3 weeks)

- [ ] Emergency credit card

- [ ] Towels

- [ ] Linens (pillows/blankets)

- [ ] Books & games

- [ ] Camp/marine shower - or watering can

- [ ] Whistle/air horn

- [ ] Rain gear/rubber boots

- [ ] Copy of key phone numbers

- [ ] Copy of essential medical information

- [ ] Extra bucket (eg outdoor toilet or shower)

- [ ] Waterproof tarps

- ☐ Clear plastic

- ☐ Battery smoke/carbon monoxide detector

- ☐ Fire extinguisher/blanket

- ☐ Solar garden lights

- ☐ Waterproof boxes (supplies/papers)

- ☐ Paper & pen for note taking

- ☐ Washing up bowl

- ☐ UV sterilisation kit

- ☐ Drinking water filtration straw (one each)

- ☐ Siphon

- ☐ Clothes line and pegs

- ☐ Disposable waterproof camera

- ☐ Sewing kit

- ☐ Non cordless telephone (for landline)

- ☐ Bicycle/construction/motorcycle helmet

- ☐ Lifejacket (with strobe light)

- ☐ Birdseed

- ☐ Candle wicks

- ☐ Sturdy shoes

- ☐ Spare socks

- ☐ Spare underwear

PET CHECKIST & SUPPLIES

The following pages contain lists of supplies and procedures which will aid you in protecting your pets before during and after a tropical cyclone. Print them off in case your devices aren't operable during or after a storm.

ALERT

☐ Ensure all pets are current on inoculations & ensure you have their papers

☐ Ensure you have an appropriate pet carrier

☐ Assemble pet disaster kit

☐ Ensure you have sufficient food & water

☐ Ensure animals have current ID

☐ Confirm animal plan

WATCH

☐ Obtain needed items

☐ Store drinking water for at least 2-3 weeks

☐ Make alternative arrangements for your pet if you plan on going to a shelter (shelters do not allow animals) – do not leave outdoors or alone

WARNING

☐ Complete preparedness activities

☐ Bring pets indoors in advance – reassure them and keep calm

POST STORM

☐ Disinfect any water they drink

☐ Walk pets on lead until they become re-oriented & be aware of insects/snakes that may come to higher ground

☐ Do not let them roam

☐ Monitor their behaviour

SUGGESTED PET SUPPLIES

☐ Proper collar/ID

☐ Proper bag/container (especially for small animals)

☐ Muzzle

☐ First Aid kit (including antibiotics, anti-inflammatories, painkillers, anti-parasite, medicine, bandages)

☐ Chlorine tablets

☐ Food (1-2 months): tinned better

☐ Drinking water for at least 2-3 weeks

☐ Can opener (manual)

☐ Rice (whole wheat)

☐ Disposable litter trays/disposable baking trays

☐ Cat litter (one month)

☐ Waste disposal bags (or latex gloves for small pets!)

☐ Paper towels

- ☐ Hand sanitiser

- ☐ Wet pipes

- ☐ Food & water bowls

- ☐ Extra dog leash

- ☐ Copies of animal's medical records (at minimum rabies & microchip certificates)

- ☐ Medication including preventative (at least 3-4 months)

- ☐ Blankets/towels

- ☐ Recent photo per pet

- ☐ Lifejacket (if in flood prone area)

- ☐ Favourite toy, treats

- ☐ Pre-made posters for each pet, or spray paint

- ☐ Pee Pads

Made in the USA
Middletown, DE
13 August 2022

71297456R00051